Alonzo Clark

Lectures on Diseases of the Heart

deliverd at the Collegr of Physicians and Surgeons New York

Alonzo Clark

Lectures on Diseases of the Heart
deliverd at the Collegr of Physicians and Surgeons New York

ISBN/EAN: 9783744718714

Printed in Europe, USA, Canada, Australia, Japan

Cover: Foto ©berggeist007 / pixelio.de

More available books at **www.hansebooks.com**

ON
DISEASES OF THE HEART

DELIVERED AT THE

COLLEGE OF PHYSICIANS AND SURGEONS, NEW YORK

BY

ALONZO CLARK, M.D., LL.D.

Emeritus Professor of the Principles and Practice of Medicine, Etc.

BERMINGHAM & COMPANY
23 UNION SQUARE, EAST | 20 KING WILLIAM STREET, STRAND
NEW YORK | LONDON
1884

CONTENTS.

LECTURE I.
Heart Sounds... 9

LECTURE II.
Pericarditis 32

LECTURE III.
Endocarditis... 96

LECTURE IV.
Myocarditis.. 116

LECTURE V.
Hypertrophy of the Heart... 122

LECTURE VI.
Dilatation of the Heart.. 132

LECTURE VII.
Fatty Degeneration of the Heart.................................. 142

LECTURE VIII.
Rupture of the Heart... 150

LECTURE IX.
Fibrous Degeneration of the Heart................................ 156

LECTURE X.
Heart Clots... 159

LECTURE XI.
Valvular Disease........ 163

LECTURE XII.
Valvular Disease (*continued*)........ 179

LECTURE XIII.
Prognosis and Treatment of Valvular Disease........ 204

LECTURE XIV.
Angina Pectoris........ 216

LECTURE XV.
Deformities of the Heart........ 222

LECTURE XVI.
Functional Diseases of the Heart........ 235

LECTURE XVII.
The Effects of Certain Drugs on the Heart........ 242

PREFACE.

THIS book contains the substance of my lectures upon "Diseases of the Heart," delivered at the College of Physicians and Surgeons for many years past, together with reports of cases collected from the literature of the subject, and from personal observation.

The basis of the work being a course of didactic lectures, the style is colloquial, and though it may not possess that rhetorical rounding which is the offspring of elaborate revision, it is given to the profession as an exposition of the views which I have inculcated during the many years in which I have been a teacher.

NEW YORK, March 21, 1884.

DISEASES OF THE HEART.

LECTURE I.

HEART SOUNDS.

I WILL first make a few preliminary remarks upon what is essential to an understanding of endocarditis and valvular diseases of the heart.

Here you observe a heart, a very large one, laid open. You observe that the inner surface of this cavity is lined by a smooth, shining membrane. This is called the endocardium, a serous membrane somewhat analogous to that which lies upon the outside of the heart. This membrane lines all these muscles, protecting the muscular structure entirely from contact with the blood. This same lining membrane is to be found in every cavity of the heart, and it is continuous in a certain way from one to the other.

Now observe how a valve is made. Here is the semilunar valve of the aorta. If you will examine this you will see that the tissue is absolutely continuous from the wall of the ventricle on to this valve. Follow it on a little further; it doubles down upon this valve, and then from there is reflected off on to the base of the aorta. Well, a valve of the heart is made, then, of a duplication of the lining membrane, of a duplication of the endocardium. The valve is closed when it lies against the aorta, and you can then hardly see how it is constructed; but open it and you see it is the same lining membrane of the ventricle which doubles down to the bottom of this valve and then passes off, without being broken, into the great vessel.

The mitral valve is made in the same way, but the lining membrane of the ventricle goes down upon the posterior face of the mitral valve, doubles upon itself, and goes back into the auricle. To this valve are attached a considerable number of little cords that go down into the ventricle and are there attached to what are called the fleshy columns. The endocardium is no less continuous for that reason. Indeed, it doubles upon these cords, and covers them. Each valve, in each part of the heart, is made in this same way, but there is, however, a little more to each of these valves. There is a fibrous structure in the middle of them—you may say a skeleton; something like the frame of a house before there are any boards or plaster put upon it. The fibrous tissue of the valve lies between two folds of membrane. Now, it happens that in endocarditis the aortic or the pulmonary valve, is most likely to show the results of inflammatory action.

But the point I want to impress upon you is the way the blood circulates through the heart, and what the heart does in the course of the circulation of the blood through it.

Suppose the blood is coming in from the lung. In the period of rest the left auricle is being distended by the blood coming in from the lung. The contraction of the heart begins in the auricle, and almost instantly spreads to the ventricle. The result of this little contraction of the auricle before the ventricle begins to contract is to send a little more blood into the ventricle, or perhaps to send a portion of it backward, because there is no valve which will prevent its going either way. Then the left ventricle contracts. What becomes of the blood? It goes into the aorta, and if there is anything to obstruct it, it may make a noise, which is not normal, and which is known as a *bruit* or murmur. Then, when you have a murmur produced at the base of the heart when the ventricle contracts it is produced by an obstruction—it may be a narrowing of the vessel at the valves. Now, what else takes place when the ventricle contracts? If there is any way for the blood to get out except the normal it will take that, as well as its natural way. Suppose the mitral valve is insufficient ; suppose it be diseased in such a way that it cannot close perfectly ; what is there to prevent the blood, when the ventricle contracts, going back over the very track it came ? Nothing in the world. Well, then, when this mitral valve is insufficient, and you get a murmur from it, it will be when the ventricle contracts : this is regurgitation. When in systole, or contraction of the ventricle, there is a murmur at the aortic opening, it is due to obstruction.

Now, suppose this aortic valve to be so far diseased, so far narrowed and shortened, that the three different portions of the valve cannot meet in the middle of the artery, and stop the blood going back ; there would be a murmur then. And when? Why, after the first sound of the heart has gone by, after the contraction of the ventricle— in the period of rest, or in the place of the second sound. A murmur, then, heard at the base of the heart in the place of the second sound, implies regurgitation ; a murmur heard during the contraction of the heart implies obstruction. You may have both in the same person, at the same time. Again, if there is vegetation, if there is unnatural thickening of the mitral valve, when the blood passes from the auricle into the ventricle during the period of repose, there may be a murmur. It is almost always a very faint murmur, and Dr. Williams described it, a good many years ago, as being like the sound produced by breathing the word *awe*—a very soft blowing murmur. That is in the period of rest; it has been called presystolic. I have no objection to the name; it is before the systole.

Dr. Flint (*Am. Jour. Med. Sci.*, April, 1882,) recognizes two very distinct tones in the presytolic murmur: one soft blowing, compared by Williams, senior, to sound produced by the breathed word *awe;* the other rough, resembling the sound produced by blowing over a piece of paper, one end of which is attached to the lower lip. He recognizes with others the fact that the soft sound is but rarely heard, while the rough sound is not very uncommon. He thinks it is produced by vibrations of the curtain of the mitral valve caused by the passage of blood from the auricle into the ventricle. "A rough presystolic murmur," he says, "exceptionally is produced when there is no mitral lesion, aortic regurgitation existing whenever the murmur is thus produced." He says that Fauvel recognized the rough murmur in 1843, and called it *bruit de rape*, and established its connection with the mitral obstructive lesion.

Then you may have two murmurs at each one of these valvular openings; in which case it is obstruction in the one instance and regurgitation in the other. Now, to satisfy yourself whether it is obstructive or regurgitative, you have only to remember the course of the blood. From the auricle it passes into the ventricle during the period of rest; then the ventricle contracts and it passes over these valves into the aorta, or if the valves are diseased the blood is regurgitated.

Then, too, it sometimes happens, as I shall explain to you further on, that these valves are subjected to a certain amount of inflammatory action when endocarditis occurs, and their action is disordered. The mitral valve, we know, is composed of fleshy columns, which are composed of fibres that are continuous with those in the walls of the heart. Now, if these be diseased, one or more of them, or a part of one may become contracted and leave an opening in the mitral valve. This fact shows that regurgitation may take place without disease of the valves, that is, without deforming disease of the valves.

As the blood comes from the ventricles it is forced over the semilunar valves, which lie smoothly up against the aorta, presenting no obstruction to the outgo of the blood. If they are rough, thickened, contracted, then they will present an obstacle to the course of the blood, and there will be produced a murmur during the contraction of the heart, that is, during systole—the time when you feel the impulse of the heart against the ear when it is applied to the chest. Now, that is easily remembered, I think; the murmur in systole, the murmur in the contraction of the ventricle, heard over that part of the chest which corresponds to the aorta, an aortic obstruction. When the blood reaches the aorta, the aorta dilates, and on account of the many little elastic fibres which it contains it has a disposition at once to contract; and in contracting it of course starts the current backward toward the heart, which instantly closes these valves and thus prevents its return. But if these valves have become so thickened, so hard and inelastic that they will not close, a portion of the blood is sent back into the ventricle, which has to do its work over. This produces a murmur which occurs during the period of repose or in the place of the second sound. This constitutes aortic regurgitation. We often have obstruction and regurgitation in the same person, so that we will have a "*whewing*" sound during contraction, which is repeated during diastole of the ventricle. It has come to be the fashion of late to use the terms "direct" and "indirect." I do not think the terms nearly as good as obstruction and regurgitation, for these are descriptive, while only "direct" can be regarded as descriptive in the other choice of terms.

Then the blood passes on in the general circulation, and comes back in the veins to the other side of the heart, to the right auricle. The right auricle in turn contracts after it is full, and sends its blood into the right ventricle, which, being connected with the left ventricle by the septum, is really a part of the same organ, and liable to contract at the same time; and the right and left ventricles then do contract at the same instant, and a sound produced by the passage of blood into the pulmonary artery occurs in systole. You observe that the pulmonary artery has valves corresponding with those of the aorta, and they work in the same way. They fall up against the artery, and allow the blood to flow over them easily and freely. But these, too, are sometimes found to be diseased. The valves of the right side of the heart, I say, are sometimes found to be diseased, and here is as good a place to state the fact as any, that in persons who have disease of the right side, if it is not due to laceration or breaking of the tendinous cords, you may pretty reasonably infer that the disease belonged to the period before birth. The right side of the heart is more subject than the left to disease in utero. The left side of the heart becomes diseased after birth. If you can find a murmur, then, that belongs to the pulmonary valves, you will be pretty sure to find it in a child, and it will be pretty safe to infer that there was endocarditis in that child before it was born. It is not a common occurrence, and yet it has happened a few times in my experience. Then the same law applies to the current of blood and to the sound that will be produced by disease of the valves of the right side as on the left. There are two valves here, the tricuspid, which guard the opening between the auricle and ventricle, and act in the same way as the mitral valve on the left side, and the pulmonary valve.

Now, I will repeat the formula by which you can recognize disease at different points of the heart. Remember that these fleshy columns are connected with the heart at a distance of from one inch to an inch and a half above its apex; these tendinous cords run into the heart, and may be regarded as conducting media, and the fleshy columns also may be regarded as conducting media. The point where, physiologically, in audition, you can get nearest to the heart, is not at the apex exactly, but about an inch above. There you will be more likely to hear sounds produced in the mitral valve than at any other part of the chest; and when I say "heard at the apex," I mean at the apex, or an inch above. When, then, you have a murmur during the time the ventricle is being filled, you ascertain, by listening at different points, where it is most distinct. If it is most distinct at or near the apex in systole, it is a murmur of regurgitation at the mitral valve; if, on the contrary, it is heard in the period of repose, the blood is flowing from the auricle into the ventricle, and that will be an obstructive sound in the mitral. The mitral is roughened, and throws the blood or itself into sonorous vibrations, as the blood flows over it; that, however, is almost always a very feeble sound. As I said before, Dr. Williams described it by breathing the word "*awe*," opening the throat well and breathing a little loud.

A murmur heard most distinctly on the third rib, at the sternum, and heard in systole, must be obstruction at the aortic valve. Then, if it is repeated immediately afterward, in the interval, you can hardly appreciate it, and yet you do appreciate that there are two sounds— the latter is regurgitation; the valve acts insufficiently, it cannot perform its office, and thus the murmurs are produced. Every little while you get two murmurs. I have heard four in the same person, two produced at the aortic valve, two at the mitral.

Now, the same law applies to the right side of the heart. You listen to the right side of the heart, at the lower part of the precordial region on the right side. I do not say the lower part of the sternum extends below the heart region; but on the sternum at the fifth rib, and outside the sternum a little, on the same level, you will get a sound from the right heart more distinctly than anywhere else. If, then, you get a murmur in that region, heard most distinctly in contraction of the heart, in systole, that would be regurgitation, at the tricuspid valve. If, on the contrary, you get a murmur which you are not quite satisfied is aortic, and it is a little to the left of the sternum, you may, perhaps, conjecture that it belongs to the pulmonary artery. The pulmonary artery is given off a little further to the left than the aorta, and you listen for the aortic murmurs on the middle of the sternum and a little to the right; for the pulmonary murmurs on the left edge of the sternum and a little to the left of the sternum. If you get a murmur there in systole it is due to obstruction at the pulmonary valve. If it is repeated, it is then regurgitation at the same valve, and is heard most distinctly at the base of the heart. But if it is a tricuspid murmur you will hear it most distinctly over the right side of the heart, at the right edge of the sternum; I never have heard a direct or obstructive murmur at this point. Indeed, a good many years ago Dr. King thought he demonstrated, and probably he did, that this particular part of the heart, the opening of the tricuspid valve, has considerable power of dilatation without injury, and when the veins are overcrowded with blood, surcharging the left ventricle, he claims that the right ventricle has the power of sending it back into the veins again through the tricuspid opening and without a murmur. I can easily comprehend that it may be so. It is what he called the safety-valve arrangement. The right ventricle being over-distended the right auriculo-ventricular opening would be dilated also, and a part of the blood can be thrown back into the veins of the general system without being compelled to go forward into the arteries that lead to the lungs.

With reference to the tones of these murmurs. You hear them in different stages of cardiac disease; mildly when the valves begin to be deteriorated; more strongly when the deformity has become considerable. And then again, perhaps not at all after the disease has continued for a length of time, and the heart has become weakened by it. Almost every person who has cardiac disease which finally terminates fatally reaches a period, before death, when the murmur cannot be heard, and the simple reason is that the heart cannot contract upon its contents with sufficient force to make the blood run swiftly enough to cause a murmur. It is probable that these murmurs are made in the solid tissue and not in the blood itself. The vibrations that produce the murmur are probably in these deformed valves. You have a great many terms by which to designate these murmurs, according to their tone. The French have used terms which represent a filing sound, a rasping sound, a sawing sound, which, of course, are harsh, and given to distinguish the harsh sounds from those which are soft and blowing. But it is not necessary for you to learn these terms. The coarse sounds are almost always produced by deformity or defect of the valves. The blowing sounds are often produced by the same thing, but they are also produced under circumstances in which the valves are not diseased; when, for example, the blood becomes watery and thin, by a law which I cannot explain to you, the blood makes a noise in going over the aortic valves. It is the anæmic sound; and, therefore, if you hear a blowing sound at the aortic opening, as if it were obstructive, your next business is to look in the patient's face and see if she is anæmic—for anæmia occurs more frequently in women than in men. If the face is pale and has a chlorotic look, you will be very likely to get a soft murmur—a blowing murmur at the aortic opening, and it will not authorize you to infer that there is any disease of the aortic valves under such circumstances.

Dr. G. W. Balfour* has lately published an article entitled "Arguments in Favor of the Theory of Dilatation of the Heart as the Cause of Cardiac Hæmic Murmurs and of the Appendix Auriculi Sinistri being the Primary Seat of this Murmur." So far as I can judge from the analysis of the abstract which I have seen (for I have not read the whole paper as published in the *Brit. Med. Jour.*, Aug. 26, 1882), it is arguments, not demonstration. He does not even adduce convincing proofs from auscultation. The idea has met with criticism already. We are not, then, converted wholly from the old doctrine that the hæmic or chlorotic murmur is produced at the valves of the heart. That in the same state of the system, the pressure of the stethescope on a vein, the jugular, for example, will produce a continuous murmur (*bruit de diable*) would seem to show that, in chlorosis, the blood changes have given to that fluid the capacity of producing sound, from causes that do not operate in that way in health. These sounds are often called anæmic, but immediately after a large loss of blood they do not occur, but they are found, if at all, after the vessels are replenished by absorption of fluid, and its constitution thereby greatly changed. This has

---

* *Medical Record*, Nov. 18, 1882, p. 572, which see, and N. Y. *Med. Jour.*, Dec. 1882, p. 657.

been denominated hydræmia, and the name is not an improper one. But the term chlorosis is applied almost exclusively to a condition that is not preceded by loss of blood, but by bad assimilation, and ends in great reduction of the nourishing power of the blood; in paleness and a puffy condition of the skin, with moderate œdema and loss of force. This is the disease in which we most frequently hear the hæmic anæmic sound.

There is another thing connected with thin blood which may perhaps help you to distinguish the sounds that are produced by deformity of the valves and those that occur when the valves are sound. Jan. 21, 1883, a boy from M——, 8 years of age, was brought to me by his parents. He had been very much reduced by disease, so that he was very pale and thin and had aortic systolic murmurs, and his physician found serious heart disease. From his anæmic state he had rapidly improved, so that now he looked healthy and was strong, appetite good, almost voracious, and digestion good. Examining him to-day there is no murmur, the heart is of proper size for a boy of his age. The inference is conclusive that the late systolic aortic murmur was a blood (hæmatic) murmur, which his improved health has removed. In most of these persons in whom anæmic murmurs occur there will be heard, through the stethoscope applied over the jugular vein in the neck, a continuous humming noise. They call it *bruit de la toupie*, which a friend of mine has interpreted with a little liberty, "devil of a noise." This *bruit de la toupie* is not the work of the devil. The French have a little toy which children whirl around, and as it goes around, a spring works on a wheel of cogs and makes a repetition of clicks. That is the devil that this is compared with—*bruit de la toupie*. This sound is not connected with disease of the heart, but as it so frequently occurs in anæmic persons it is worth while to allude to it in this connection. The harsh, rasping sound is almost always produced by some defect of the valves; the soft, blowing murmur may be so produced, but it is occasionally found in persons whose hearts are sound enough, but whose blood is diseased or in a watery condition. I do not think there is any practical use in making any other distinctions between the soft and the harsh, or the blowing and the rough murmur.

Dr. E. Hyler Graves* concludes: 1. The presence of murmurs does not necessarily indicate the existence of incurable lesions, or their absence that such lesions are not present; but other symptoms must be looked to for a correct diagnosis. 2. The persystolic murmur of mitral stenosis may disappear and the lesion remain. Mitral regurgitant, when due to simple relaxation of the heart's muscles and dilatation of its cavities and orifices, as in chlorosis and fever, in most cases completely disappears. 3. Tricuspid regurgitation is occasionally temporary, due to bronchitis, etc. 4. Aortic systolic murmurs, due to permanent lesion, may undergo changes in intensity but never completely disappear. 5. Aortic diastolic murmurs in rare cases have been known to disappear. The systolic aortic murmur in these cases is always present and is persistent. 6. The pulmonary systolic is permanent when due to organic lesion, but is non-organic; may disappear temporarily or permanently.

There is another thing that I must call your attention to before discussing endocarditis itself.

At the 83rd meeting of the *Medical and Chirurgical Faculty of Maryland*, April, 1881,† Dr. John S. Lynch communicated a new plan in the differential diagnosis of cardiac and pericardial murmurs. He said, "Whenever the friction murmur is produced at or near the apex of the

---

* *Med. Rec.*, Sep. 29, 1883.
† *Am. Jour. of Med. Sci.*, Oct., 1882.

heart, if we cause the patient gradually and slowly, but entirely to inflate the lungs, we will perceive that the friction murmur becomes gradually progressively more intense until the act of insufflation is complete. Now make the patient 'hold his breath' while the lungs are in this state, and the murmur will be steadily maintained at its maximum intensity; cause him then to expire in a like slow and gradual manner," and the reverse will be observed. He thinks the sound at the apex is the only one that presents any difficulty in diagnosis.

The heart itself has two sounds. They are denominated first and second, and it will interest you to know what is the cause of these two sounds; and, perhaps, as great diversity of opinion has been expressed in regard to this, it may be of some little interest to you to know which opinion I have chosen.

As a consequence of the arrangement of the muscular fibres of the heart, as the heart contracts, all these bundles of fibres contract at the same time, and one rubs against another. It is not exactly a fiddle-string operation, and yet it is not without analogy to it. These fibres cannot rub upon one another actively, as they do when the heart contracts, without producing some sound, and this first sound continues during the contraction of the heart. It begins with the beginning and ends with the end of that contraction—that is, of the ventricles. The auricles are placed a little out of hearing, so that you cannot know what is going on in them by audition. You can imitate the sound very closely by placing the stethoscope upon the ball of the thumb and opening the thumb. You will hear a sound exactly like that which is produced by the first sound of the heart. Another very good illustration of this sound may be had by placing the arm against something that is firm and extending and contracting it, and listening to the action of the muscles.

It has been said that the first sound is produced by the tension of the mitral valves, in consequence of the sudden jerk made upon them by their muscular columns. I do not know that that does not enter into the first sound and make a part of it; I do not know that it does. It has been said, again, that the first sound is produced by the rush of blood from the auricle into the ventricle (?). I do not know but that this aids in the production of the first sound, and I do not know that it does. The current is easy, the flow is at a pretty good rate, to be sure, and I should not be surprised if concussion against the empty walls of the ventricles would produce some effect of that kind ; and yet I do not know it.

I think when you have tried these little experiments I have spoken to you of, and then listen to the heart sound, you will be pretty well satisfied that the chief element in it is the contraction of the muscles, and the rubbing of these cross-fibres one against another. They are separated, to be sure, by a little layer of connective tissue, but even that is put upon the stretch by the action of these muscles that are working at right-angular contraction. The first sound, therefore, is confined entirely to the systole, to the contraction. The second sound there are no debates about ; experiments have been performed that explain that perfectly. A number of physicians in London tortured some honest old asses and worn-out horses, by opening the chest and having little hooks that were sharp enough to penetrate the tissue, laying the hook-handle along the course of the aorta and penetrating it, then carrying the hook forward and getting it over one of the folds of the semilunar valve, and drawing it back up to the wall of the artery. It is an experiment that can be pretty easily done. The result of the experiment was a cessation of the second sound. Then by depressing the handle and raising the point of this hook the valve would be let loose, and be at liberty to play again, and then the second sound was reproduced. They repeated the experiment a great many times, and in a great variety of ways. A Philadelphia committee of doctors repeated it with the same result, so that we have the fact pretty well established, that the second sound, which you will recognize as instantaneous as a click, results from the falling together of these three segments of the valve in the centre of the aorta.

When the second sound of the heart is unusually loud, it has been the habit of late to say that it is *accentuated*. This word may not be the best that could have been chosen, but it appears to have the sanction of usage, and that entitles it to the place it occupies. This intensified second sound may be noticed on either side of the heart, that is, it may be seated either in the pulmonary or in the aortic valve. In either, two conditions of the organ are necessary. It must retain a good deal of its natural strength, and the valves referred to must be sound, or if not entirely free from disease, not diseased in a way to impair their efficiency. Another condition is equally necessary—that the flow of blood from the half of the heart in which it is produced be resisted. The familiar second sound in the pulmonary artery, when there is stenosis of the mitral is an example. Any obstruction in the pulmonary circulation, the heart retaining its strength, the valve its integrity, and the quantity of blood in the system undiminished, will produce it in the same way. Pneumonia produces it unless the heart is feeble. Phthisis does not, because the volume of blood in the body is diminished, as is also the strength of the heart.

Dr. Begbie has found that sacculated aneurism and dilatation, or true aneurism of the aorta, produce, with a good deal of constancy, exaggerated second sound at the aortic opening. The probable explanation of this is that as neither of these changes occurs spontaneously without atheromatous deposit in the vessel, that while the strength of its walls is diminished, their distensibility is also decreased.

You would naturally expect, if these aortic valves were inefficient, and did not guard the opening, that there would be no second sound produced. But call to mind the fact that I just now stated to you that the two hearts beat at the same time, and it is almost never found to be the case that the pulmonary valves and the aortic valves are diseased in the same person. You have, then, the pulmonary valves to give you the second sound. The aortic valves may not participate in it, may not fall together at all; they may be too short and too inflexible. The cause, then, of the two sounds of the heart you can easily carry in mind. The reflux of the blood from the aorta in its attempt to get back into the ventricles, closes, as I told you, these valves; they close with a click, and that is all there is of the second sound.

The *sphygmograph* is now pretty well known, and its indications pretty well understood. It will indicate aortic obstruction by the sloping rise of its tracing; aortic regurgitation by the rapid fall; irregular action of the heart by its writing down the small and varying action with more correctness than the finger on the artery can possibly attain; and yet its diagnostic value is not great, for almost all that it teaches that is important can be learned by other and shorter methods.

But now there is a compound sphygmograph, by which the beating of the pulse of any artery, and the beating of the heart are recorded on the same paper, one tracing over the other with a time index at the bottom. It shows how much the pulse wave in any artery chosen for the experiment is delayed, after the heart has contracted, or on the other hand, if the pulse wave reaches the artery under examination in less time than the normal, how much it is accelerated. As each contraction of the heart and the pulse produced by it is traced one above the other, it is easy to compare them, and to learn the time difference between them. But that this difference may be useful, the natural or normal difference between the heart beat and the pulse beat in the several arteries must be known. This basis fact is apparently ascertained. Dr. A. T. Keyt announces that the time consumed by the pulse wave in passing to the carotid artery is $\frac{1}{12}$ of a second; to the temporal in front of the auditory meatus, $\frac{1}{10}$ of a second; to the radial, $\frac{1}{8}$; to the femoral, $\frac{1}{7}$; to the artery on the upper surface of the foot, $\frac{1}{6}$. These are averages. If they are confirmed they make an interesting demonstration of the rate at which the blood travels in the vessels. It is claimed that dilatable aneurism, stiff aortic valves and mitral regurgitation delay the wave, and make the time longer; while aortic regurgitation and rigid arteries make it shorter. Dr. A. B. Isham of Cincinnati has published a paper on the application of this instrument to diagnosis (*Am. Journ. of Med. Science*, July, 1882,) in which he gives a considerable number of cardiac and arterial tracings, one over the other. The greatest difference that he has recorded is $\frac{28}{88}$ of a second between the heart and the right subclavian, while in health the time should be $\frac{5}{26}$. This is certainly a marked retardation—one that could be recognized but not measured without the aid of the instrument. It occurred in a person who had aneurism of the aorta and mitral regurgitation. In a case of mitral regurgitation the pulse in the carotid (right?) was delayed $\frac{1}{8}$ to $\frac{1}{9}$, instead of $\frac{1}{17}$ of a second. But these are large delays. In a case in which the aortic opening was greatly obstructed, the delay in the carotid (right?) was $\frac{1}{4}$ of a second, while the natural delay is $\frac{1}{12}$ of a second. This case is narrated by Dr. A. T.

Keyt, in the Cincinnati *Clinic*, April 19, 1879, and in the *Medical Record*, June 4, 1881. I have been sufficiently interested in the use of this instrument to try to obtain one, but it must be imported, and my confidence in its practical value has not yet overcome this trifling obstacle; for so far its teachings are more confirmatory of what can be learned by other means, than an instrument communicating new information. But as the shrub may grow into a tree, the moderate beginning of the cardiac sphygmograph may have a large and expanded growth.

Heart Scanning.—Dr. Samuel W. Francis says that the normal beat of the healthy heart is iambic ⌣ —; that when it is trochaic — ⌣, pyrrhic ⌣ ⌣ or spondaic — —, these measures indicate disease. He also reports a case in which the pulse was only 29 in the minute, and the beat dactylic — ⌣ ⌣, a long and two short being well marked. The patient was a lady sixty years of age, who recovered under diffusive stimulants, and counter-irritation. (*The Medical Record*, Feb. 24, 1883.)

Dr. Henry Cook has quoted three cases published by Dr. Hyde Salter, and added three of his own, in which the rhythm of the heart was lost in a double beat, one following the other so quickly, that there was hardly time enough between them to perceive that there were two— the end of one instantly followed by the beginning of the other. The second systole was followed by a second sound, but the first was not. Thus,

___ ___ -   ___ ___ -   ___ ___ -
                      instead of
___ -   ___ -   ___ -   ___ -   ___ -   ___ -

In these cases the first of the double beat sent a pulse to the wrist, the other did not; but in one of Dr. Salter's cases this was changed. At the first examination it was the first of the two beats that produced the radial pulse;

at another, he says "the second beat of each couple was the strongest, not the first as formerly, and the only one reaching the wrist, though the first and feebler was plainly visible in the carotids."

In four of these six cases it was noticed that the time occupied by this double beat and the pause following, was just double that required for a single beat and pause. For example, if, while the heart had a single beat and pause, the pulses were 106, when the double beats occurred it would be 53. Both observers had the opportunity of listening to the heart at the moment the change from double to single beat occurred, and Dr. Cook witnessed the change from single to double. The change in each direction was not attended with the slightest disturbance; indeed of one patient it is said he did not know that any change had occurred.

In a man who had aortic obstruction, and both tricuspid and mitral regurgitation, Dr. Cook says, "each impulse was accompanied with the systolic bruit." He does say that there was disease of the pulmonary valve, and no post-mortem examination is reported.

These double beats were strong, each a good deal stronger than the single beat, though not always of equal strength. Neither of these gentlemen entertained any doubt that it was a double action of the left ventricle, and Dr. Cook devotes considerable space to conjectural explanations of the mode in which it was effected, none of which seem to himself wholly satisfactory. Was it not that the beat of the right heart followed that of the left when the radial pulse was produced by the first contraction, and opposite when the pulse attended the second? But can the action of the two hearts be partly separated in that way? Some recent experiments referred to elsewhere seem to prove that it can be, in animals, and to a much greater extent than is necessary to explain Drs. Salter and Cook's cases. But I have myself, not frequently, to be sure, but two or three times recognized a double second sound, one following the other quickly, but each perfectly distinct, and have called the attention of persons who were with me to the fact as showing a lack of strict coincidence in the action of the two hearts. Dr. Cook has given sphygmographic tracings of his cases, which appear to be conclusive on this point. If the two beats were both left ventricle contractions, when the tracings reached the highest points, they must be held there during all the time of the second beat of the couple, with perhaps a downward notch between the two. But there is nothing of the sort. The descending line pays no attention to the second beat, but descends at the acutest of angles, greatly more acute than when the tracing is of the normal and regular beats.

The only facts that seem inconsistent with this explanation are, first, the statement that Dr. Salter noticed a pulse in the carotids when it did not reach the radial artery. But may not a full wave in the heart give a pulse to the aorta and pulmonary artery that would extend to the carotids? The two ventricles so commonly contract at the same instant that observations are yet to be made which will enable us to answer this question positively. But the affirmative answer seems to me probable. Second, Dr. Cook says that in the case in which all the valves were diseased except the pulmonary " There were now two impulse beats, then a second sound, and then a long pause, and each impulse was accompanied with a systolic bruit." If this statement is strictly correct, there must have been disease of the pulmonary valve also. At least it is easier to suppose this than to admit that a ventricle just emptied of blood could empty itself again on the instant, without time to get a new supply, and further, with no power to make a pulse.—*New York Medical Abstract*, Feb., 1882.

Cardiac Irregularities.—Dr. Lukjanow, not long ago, tried the effect of closing one, and then the other of the coronary arteries of the heart; and asphyxia on rabbits and dogs. The functional connection of the different parts of the heart was interrupted by the closure of one coronary artery. A difference in the number of contractions of the auricles and ventricles was easily produced, and almost as easily a difference between the two auricles. " Asynchronism is produced in the auricles much more readily than in the ventricles." The closure of one coronary is found to influence, first, the ventricle on the same side; then the other ventricle, and lastly the auricles." The cardiac muscle may contract, he says, in paristalsis, or in anti-peristalsis. Sudden asphyxia especially effects sometimes one, sometimes the other side of the heart. He attributes these phenomena to "the effect of sudden ischæmia, and to the retention in the tissues of the products of such action as may occur."—*Am. Jour. of Med. Sci.*, Apr., 1882.

If these experiments can be relied on, asynchronism can be produced by local anæmia and by asphyxia, and if by these causes in all probability by others not yet understood. There is no chance closure of either coronary artery in either of these patients, but the circulation may have been modified by atheromatous or calcareous deposits in one of them, or, as is most probable the asynchronism came from some unexplained condition of the nerves of the heart.

As to these heart nerves Dr. E. P. Hurd (*The Medical Record*, Oct. 28, 1883,) writing of the Physiology of the Heart, says the cardiac innervation comprehends certain intra-cardiac ganglia, extra-cardiac ganglia and nerves derived from the sympathetic and cerebro-spinal system.

Intra-cardiac Ganglia.—These are in the walls of the heart, and regulate the rhythmical working of the cardiac muscle. This is a property of the myocardium itself also. The ganglia of Remak are placed at the point where the sympathetic and pneumogastric nerves enter the heart, whose terminal filaments are lost in these ganglia. They occupy the sinus of the vena cava. Other ganglia, those of Bidder and Ludwig, are situated in the auriculo-ventricular furrow. In these ganglia originate fibres, some of which, centrifugal, are distributed to the muscular fasciculi, others, centripetal, terminate in the endo-cardium. These ganglia and these nerves constitute the excito-motor arc, and are to a certain extent independent nerve centres.

Extra-cardiac Ganglia.—Almost every nerve centre has some connection with the heart, but the cardiac centre emphatically is in the rachidian bulb between the tubercula quadragemina and the thalamus. When this is stimulated by a strong electrical current the heart's action is arrested or slowed by a less powerful current. If previously the pneumogastric nerves are cut excitation here produces cardiac acceleration. The medulla oblongata is especially the centre of impressions which reflexly affect the frequency and rhythm of the cardiac movements.

Other extra-cardiac ganglia exist in the cervical portion of the spinal cord. These may be regarded as the centres of the accelerator nerves. Von Bezold has shown that excitation of the spinal cord, especially in its upper portion, augments the energy of the heart and the arterial pressure. These auxiliary cardiac ganglia are in the closest relation with the vaso motor centre, and any excitation which raises the arterial tension may be attended by accelerated cardiac action. Ludwig and Thery have clearly proved this intimate relation.

The depressor nerve of Cyon is a sensory nerve found among the filaments of the pneumogastric, coming with it from the medulla oblongata. An excitation of its terminal branches, endo-cardial, is reflected on the splanchnic nerves opens the splanchnic vessels, so that the heart work is lessened and the blood pressure lower.

The vaso-motor nerves, when they contract the vessels, make more work for the heart and increase its beatings. This effect is witnessed when the splanchnic nerve is excited by electricity, the blood pressure is increased and the heart beats are quickened, and contrarywise.

There are a few points in reference to the anatomy of the heart which are necessary to the clear understanding of the diseases of the organ, and these we shall proceed to touch upon.

Other investigators report observations on the blood pressure in the coronary artery and the carotid. Simultaneous tracings taken in a branch of the left coronary artery, and in a carotid agree in every respect. If the aortic valve closes the coronaries during systole, the period of greatest pressure in the arteries of the heart should follow the same period in the carotid. When one inspects the aortic valves (*N. Y. Med. Jour.*, March 3, 1883,) lays them up against the aorta, and sees how completely they cover the openings of the coronary arteries, and must cover them when the valve is open, it is much easier to believe that these gentlemen have made some mistake in their experiment, or in their explanation of it, than to abandon the apparently inevitable inference from this piece of anatomy. But if the experiment is correctly reported, does it prove any more than that the blood pressure is greatest *in* the coronary arteries, when the muscular pressure is greatest *on* them (in systole), for at the moment of greatest pressure their mouths are closed and held closed by a force equivalent to that very pressure.

Mr. Geo. C. Karop says (*N. Y. Med. Jour.*, July 7, 1883) that while the best authorities are divided, some asserting, some denying that the coronary arteries anastomose, when junior Demonstrator of Anatomy at Middlesex Hospital, he made some injections, and found that in some cases they anastomosed, and in others they did not, the latter being as two to one.

The question requires further attention, "but the conclusion will probably remain as before, that in some cases the anastomosis is very free, in some slight, and in others does not occur at all."

LECTURE II.

PERICARDITIS.

THE pericardium is the sac or bag *about the heart*, containing it. It is double in two senses. The real sac is made up of two structures or tissues. The outside one is formed of fibrous tissue, not unlike that which forms the ligaments of the joints, but less condensed, and is continuous with cordiform tendon of the diaphragm below, and the outer covering of the large vessels above. The inner layer is a serous membrane, no more and no less incorporated by interchange of fibres with the outer layer, so that it cannot be separated by the scalpel without considerable labor. These two constitute the *heart sac*. From this, at or near the origin of the great vessels, the inner or serous layer is continued on to the heart and wholly invests it, forming a closely adhesive layer. Serous membranes everywhere are constructed on the same plan, lining the parietes and covering the viscus without break of continuity.

The term pericarditis is used to designate the inflammation of this serous lining of the fibrous pericardium and this serous covering of the heart.

Here, when we have to use percussion as a means of diagnosis, is as good a place as any to give an explanation of a method by which the auscultation of percussion is in certain respects improved — into which a new principle is introduced. The percussion of Laennec gives information regarding the condition of the different organs of the body by the greater or less volume of sound, and by modification in the tone of sound produced by it. This results from vibrations produced in the covering structures, and will be freer and of deeper note if air or gas or a rarefied structure be underneath, but of higher pitch and scanty volume if a solid body or fluid be in contact with the percussed surface. Every carpenter in the land understands the practice growing out of this principle and applies it when he is to drive a nail into the plastered wall of a room. He taps with a light hammer on the wall from side to side till he strikes a spot where he finds the sound is not "hollow" and is of less volume, and he knows that he has found a timber that "will hold his nail." Persons keeping fluids in wooden casks apply the same principle to ascertain how much fluid is left in them. The farmer strikes the cider barrel with his knuckle, from above downward. At first the sound produced is clear and free, and soon it is suddenly changed in volume and tone. He knows that is the cider level. The sound in all these cases is produced in the containing wall.

In the year 1840, probably in January, the late Dr. Cammann called on me with a new idea. It was no more than a thought. He had not attempted to prove or apply it. May not vibrations be produced in solid or thick walled organs which can be brought to the ear as sound? He had no sooner stated his point than I saw, or thought I saw, a new field of useful investigation. We agreed at once to give our leisure time to the inquiry. We met day after day at the dead house of Bellevue Hospital, where I was already an accepted volunteer, and with stethoscopes of solid cedar wood, one inch in diameter, and six inches long, or wedge-shape, or almost pointed at the objective extremity, each furnished with an ivory ear-rest through which the cedar projected slightly, we began and prosecuted the study. Dr. C. S. Mitchell, then of New York, now of Brooklyn, often met with us and gave us his assistance. The mode of proceeding was this: The objective end of the stethoscope was in central position over the heart, for example, or over the spot where the lung does not overlap it, and the ear was applied to the aural end, leaving the two hands free. The forefinger of left hand was the pleximeter, and the first two of the right the hammer. By standing on the right side of the body, and then on the left, we could easily percuss in the whole circumference of the heart. We percussed first off the heart and made the percussion approach it in a right line on every side in succession. While the percussion had not reached the organ, the sound was distant and of small volume; but when it reached the outer border of it, on either hand, the sound grew instantly louder and acquired a half metallic ring. This point was accurately marked. In this way the whole circumference was mapped out, and then sharpened knitting needles were driven into the chest, at these marked spots, perpendicular to the plane of the body. We were surprised at the accuracy of our measurements. The needles were always in contact with the pericardium, often entering it without wounding the heart. These experiments were repeated day after day by ourselves and by friends, and always with the same results. Upon the dead body, then, the demonstration was complete.

We tried the same method for fixing the boundaries of the liver, the spleen, and the kidneys, and found it equally accurate on them, with this exception, that the liver was the least conductive of them all, but that the limits of this could be easily defined when the stethoscope was not more than two inches and a half from the border.

These facts and others of the same kind were arranged for publication by myself, and appeared in the July number, 1840, of a journal edited by Drs. Watson and Swift.

Soon after this paper appeared, a meeting of the New York Medical and Surgical Society was held in the Library of the New York Hospital, and a patient was brought in on whose chest the boundaries of the heart had been marked by an ink line, and the members, one after another, tried the new method, and, as I now remember, each member indicated the same line, some at once, some after a second trial and some hesitation. I state this to show that the art of this plan is not difficult of acquisition.

My lamented friend Dr. Powers, of Baltimore, soon after reading the paper, was called to see a patient who carried a large abdominal tumor, the nature and relations of which were not determined. He could not himself percuss anteriorly and listen behind; another percussed for him, and he carried the stethoscope from near the point of percussion backward to the usual place of the kidney of that side, and finding no break of continuity, only a uniformly diminishing volume, he pronounced the kidney and the tumor continuous structures. The size of the tumor rendered this opinion improbable, but inspection after death proved it to be correct.

Not long ago a gentleman was brought to me from the

West by his physician, who was not satisfied with the term *malignant* which had been applied by a surgeon to an apparent tumor in the right side of the abdomen, and much less with the surgeon's proposition to remove it. I found it movable, and especially that it could be pressed upward and backward toward the place the kidney should occupy, but particularly on placing the patient in a horizontal position, his elbows on a chair. I found, by auscultatory percussion, that there was no organ where the kidney should be to give any sudden change, at its borders, to the percussion sound. On these two facts was based the opinion that the "tumor" was a floating kidney. I am not informed, and may never be, whether this opinion was correct.

Last winter a large consultation of the Medical Board of St. Luke's Hospital was called to decide on the propriety of removing what appeared to be a diseased kidney. Auscultatory percussion was tried in that case, the stethoscope being placed over the kidney and percussion made from it toward the right (it was the left kidney), and it was found that the peculiar percussion note ceased where the right limit of the kidney should be, and was not renewed when percussion was made on the tumor. It was, therefore, stated that the kidney had no structural connection with the tumor, and the hæmaturia which had been a misguiding symptom might come from the extension of the malignant disease of the tumor to the ureter. This opinion, suggested mainly by the stethoscope and percussion combined, was proved to be minutely correct at the examination after death.

Auscultatory percussion may be practised on all organs to which it is applicable, in the same way as it was on this kidney. The information given by losing the percussion note is of the same import as that given by finding it, and gives equally reliable information regarding the limits of organs. This is true even of the liver if the stethoscope is not placed too far from the border to be examined.

There is another mode of using the stethoscope and percussion together which is explained on a similar principle. The instrument may be placed entirely off from the organ under examination, and at a distance of two or three inches from it, and percussion made first on the organ and toward the instrument. When the percussion passes off the organ a new and sharper note is heard. This is more true of the chest than of other parts, because the walls of the chest are more fixed and more conductive. For example, the percussion sound from the heart when the instrument is placed an inch or two from it is dull and distant, and as the percussion passes off its border it becomes sharp and clear. The percussion vibrations are taken by the muscles and tissues of the chest walls, while, so far as the heart extends, these vibrations are more or less stifled and smothered.

I have made this digression in favor of auscultatory percussion, because I do not believe the knowledge of that mode of examination is at all general in the profession, although I have explained it every year to medical classes since the paper was published. The journal in which it appeared did not survive its first year, killed outright by a review from the pen of one of its editors, and has been seen no more. I have been told, however, that it received, at the time, the distinction of being translated into another language and published in an undistinguished medical journal, the authors' names suppressed and the translator's name appearing in their place. I have not seen this translation and theft; my informant said that he had. Any way it appears that there are few either here or abroad that know anything about the matter; perhaps none now practice it except those who have listened to my lectures.

As to the form of stethoscope, that already described is probably the best, but any stethoscope will answer the purpose. I have often gone back to the primal instrument of Laennec, a rolled pamphlet, when a better was not at hand.

In its application to the heart of a living person the indications are not always as satisfactory as in the dead body. The reasons for this are—1st, That in the latter the lungs are in the position of rest, *i.e.*, expiration; and 2nd, that the depth of lung overlying the heart in the living is constantly changing, in inspiration more, in expiration less; and then the depth of lung over the heart varies in different persons, and may make the exploration easy in one and difficult in another. In case of difficulty it will be well to have the patient make a full expiration and suspend the breathing for a few seconds while the physician percusses; or the latter may place the instrument an inch or more away from the probable place of the border of the organ and percuss toward it from the heart.

There is another point to be made: the ribs are excellent conductors of sound, much better than the tissues of the intercostal spaces; if, therefore, the instrument be placed on a rib or cartilage and percussion is made on the same rib, the sound may seem to indicate a much greater size of the heart than is real. If, then, the instrument rests on a rib the percussion should be made above or below that rib, or the intercostal stethoscope may be used, which is wedge shape, and applied so as not to touch any rib. The stethoscope that tapers down from an inch in diameter at its aural extremity to $\frac{1}{8}$ of an inch at the applied end will serve equally well, although this was devised for the purpose of detecting small deposits of solid structure or thickened tissue anywhere in the body.

With these cautions and helps it will be in only a very exceptional case that the observer cannot make out the position and boundaries of the heart with accuracy. He will, of course, confirm the indications given by the stethoscope by those discoverable from ordinary percussion and the point of the apex beat.

The result of this work thus detailed, so far as the heart is concerned, was to show that its upper boundary was the upper border of the left third rib, and this line extended across the sternum, while the left auricle rose into the second left intercostal space when filled; and the right, extending half an inch from the sternum to the right, when in a like condition—that the left boundary was a curved line beginning on the third rib three inches from the median line—on the fourth rib four inches, and in the fifth intercostal space $3\frac{1}{2}$ inches.

Morbid Anatomy.—In acute pericarditis there is doubtless the same congestion of the blood-vessels that is observed in the first stage of other serous inflammations, though little of it is seen when the disease has run its course. There is, in all probability, in all cases, some production of pus, though commonly not enough to be seen by the unaided eye. This seems to be a common *event* in serous inflammations and perhaps in all inflammations, and is regarded as due to the transition of the white corpuscles of the blood through the walls of capillary vessels. Cohnheim first saw these bodies working their way through, and appearing on the outside of these vessels, and thought he recognized their identity with the pus corpuscles. This observation has been confirmed by many workers with the microscope.

The red corpuscles of the blood sometimes break bounds in the same way and stain the other effusions.

But this does not occur in visible quantity except when pericarditis occurs in persons already affected with scurvy, or in persons whose condition is allied to that prevailing in that disease. Then the quantity of blood may be considerable and the disease is called *hemorrhagic* pericarditis. In this condition if persons are attacked by this disease or by pleurisy, it is pretty safe to infer that it is hemorrhagic. A prisoner for some misconduct was put on a diet of bread and water. This discipline diet was continued for some weeks, and the man was much reduced by it. In this state he was brought to bed by pericarditis, and sent to the hospital. I assumed that it was the hemorrhagic form, not from any peculiar behavior of the disease, for it has no distinctive symptoms, but from the man's history and condition; and so it was. In a hundred cases occurring under ordinary circumstances no blood will be *seen*.

The common effusions are fibrine and serum. The first clinically recognizable is the fibrine. A day or two later the fluid, if it become at all abundant, can be found. The appearance of the fibrine is variable. In one person it may be spread as a thin, smooth membrane over all the inner face of the pericardial sac and the outer surface of the heart. In another it is more abundant, and appears to have caused attachment of the pericardium to the heart, but afterward these surfaces are separated by the interposition of serum, and a part of the fibrine is seen attached to one surface and a part to the other. In rare cases this division of the fibrine between the heart is pretty even, giving to each nearly an equal share. Then it may be found that many little threads of fibrine run through the fluid from one division to the other, looking like little pillars. But much more frequently the division is very unequal and is complete. Then the surfaces of the effusion may be very rough. On one side lumps of irregular shape may be left, corresponding with dents or depressions on the other, or the heart may be partly covered by little cones or nipples, and the pericardial surface be left nearly without fibrine. This latter arrangement was once thought to be significant, and had the epithet "scabrous" attached to it. But it was regarded as indicating nothing as to the nature of the inflammation. It is evident that when serum is effused after the fibrous membrane is formed it finds a place for itself, when this new membrane will most easily yield to it, and that this effusion is more firmly attached to living membrane than to itself, and is more easily split into layers than detached. The mode of separation is of little importance when it is properly understood.

The quantity of serum in acute pericarditis varies greatly. Often there is so little of it that its presence is hardly appreciable during life and appears insignificant after death, but, in most cases, it is found in the sac to amount to a few ounces or even to a pint or more.

The sequel of the plastic effusion is important if the patient recovers. During recovery it seals the pericardium firmly to the heart, so that if the patient dies months after the pericardial inflammation has passed away the adhesion is still found, though in the mean time the patient thought he was well, and acted like a well man, but the time will come if the patient live long enough when the separation will be complete. A child of six years had rheumatism and pericarditis under the care of Dr. Smith of Randolph, Vt. He grew to be a man, and was an industrious and useful man. He died at the age of twenty-eight, and here is his heart. You will see that the pericardium is attached to the heart, not in close adhesion, but by countless numbers of small threads half an inch or so in length. The spaces between the attachment of the threads show healthy pericardium. The next step will be the rupture or absorption of these threads. If it is true that "old Pan" had a "hairy heart," it must be explained by the supposition that these threads were found broken but not wholly absorbed; consequently that he had had pericarditis.

The evil caused by this adhesion can be best understood by referring again to anatomy. If you dissect a heart, say a bullock's heart that has been boiled, you will easily see the muscular-bands running in almost every direction and one layer in contact with another, and particularly you will find a band beginning at the base of the heart, and running obliquely downward and to the left till it comes near to the apex, when it makes a corkscrew turn about this part of the organ. The effect of this is to draw the apex forward with each beat. This explains the fact that when this organ is healthy we feel the apex beat easier than that of any other part of it. It implies, also, a twisting movement of the apex in the heart-sac—a movement easily performed when there are no adhesions, but resisted when the pericardium is adherent by the attachment of the latter to the cordiform tendon of the diaphragm. You would naturally infer from this that as more strength is needed the need would be supplied by more strength in the heart, in other words, by hypertrophy. But here I show a heart on which the pericardium is completely attached in every part, and had been so for twelve months before death, and there is no hypertrophy whatever. The attaching membrane is very thin and delicate. But here is one in the same condition, except that the intervening material is very thick and stiff, and does not seem to be fully organized, the new membrane having a leathery character, and must have been in itself an obstacle to the heart's action, yet even here the heart is not hypertrophied. On this tray I show you several specimens illustrating the same thing. The rule, then, seems to be that simple adhesions of the pericardium to the heart do induce enlargement of that organ. And I do not remember any exceptions. The rule is very different, as you will see by and by, when the valves of the part are implicated. .

One remark forces itself upon me here. In some of the serous inflammations it seems to be demonstrated that there are little granules of connective tissue cells studding most of the inflamed surface, and that the granules and not false membranes are the agents of attachments and adhesions, while the false membranes become fatty and are absorbed. In many of the hearts you are now examining I admit that in their present state we cannot, on them, disprove this theory, but how is it with that one in which the new material is nearly an eighth of an inch in thickness? Is it possible that adhesion is by agency of such granules? The fleecy appearance in the central layer of this shows that the false membrane is still there.

There is one thing more to be said while on the pathology of pericarditis. Here is a heart to which the pericardium is firmly attached on the left, while it is separated from the right side and greatly distended or bulged. It is from a person about twelve years of age, who for several successive years had an attack of rheumatism in the spring, attended almost every time with pericarditis. The last was attended by some œdema, great cyanosis and distressing dyspnœa. There was marked extension of pericardial dulness, especially to the right. This heart was found in the condition in which you see it, except that the distended part of the pericardium contained a large quantity of serous fluid, which so pressed upon and compressed the right heart as at length to make the circulation impossible. The inference is that when the last attack of pericarditis began the adhesions were over all the heart, and that the force of the new effusion separated the attachments so far as they are separated, and that the adhesions elsewhere were too firm to yield to it. Your own examination will show you that what remain are very firm.

In cases in which there are no previous adhesion the amount of serous effusion is very large and very distressing. This will be referred to when speaking of symptoms.

In chronic pericarditis there is always an abundant seropurulent effusion, and the pericardium is much thickened.

Causes.—Pericarditis does occur sometimes as an independent disease, but it is almost always secondary to or at least co-existent with some other disease; for example, Dr. Latham had in four years at St. Bartholomew's Hospital 136 cases of acute rheumatism, in which he found the heart affected in 90, the endocardium, however, oftener than the pericardium, *i.e.*, in 74—the pericardium in 18. Another observer gives an analysis of 1410 cases of all diseases. There were in this number 161 who had acute rheumatism. Of the 1410, 85 had recent pericarditis—61 in the rheumatic cases, 1 in 3, say, and 24 among patients that had not rheumatism. Of his 24 cases not rheumatic 7 followed inflammation of the lungs or pleura; 2, malignant disease of the heart; 1, old cardiac disease—6 were connected with granular kidney; 4 followed hemorrhage and exhaustion; 2, scarlet fever and erysipelas, and 2 were not explained by any preceding or attendant disease. Dr. John Taylor found pericarditis of a severe grade, in 1 out of every 80 cases in the University College Hospital, and of these about two thirds occurred in the progress of acute rheumatism. The connection between pericarditis and acute rheumatism was established by Bouillaud, though partly seen by several persons before him. Perhaps the first person who recognized the fact that acute rheumatism can produce "disease of the heart" was Dr. Smith of Randolph, Vt., in a case to be spoken of hereafter. In persons suffering from Bright's disease, whether it be the variety known as the large white kidney, or the contracted granular kidney, a certain number contract acute pericarditis. Dr. Latham, as I have told you, had 6 cases of this kind in 85. Dr. Taylor in his 35 cases, ascribes one third to Bright's disease.

The relations of acute pericarditis and pneumonia deserve attention. In fifty years of attendance on hospitals in one capacity or another, I have been many times surprised to find acute pericarditis in persons who were supposed to have died of pneumonia alone. This did not arise from any real difficulty in diagnosis, but from the fact that the symptoms appeared to all accounted for by the recognized pneumonia, and no search was made for any complication. The pericarditis is as easily recognized in this connection as under any circumstances. After several such surprises I adopted the practice of listening for the heart disease in every case of pnenmonia. Dr. Latham had 7 concurrent with this disease and pleurisy together.

As to the relations of acute pericarditis and scarlet fever, I do not remember any instance in which the heart disease has occurred in the stage of eruption, but, with me, it has occurred several times in the œdema that is a frequent sequel of it.

Pericarditis may be caused by inflammation beginning in neighboring tissues, and by abscess of such parts opening into it. In the latter case the inflammatory effusion is pus and fibrine. So it is if the disease occur in the course of pyæmia. A singular case of purulent pericarditis will be narrated by and by, artificial teeth and the plate on which they were set having passed into the œsophagus.

The white hard patches so often found on the anterior face of the heart, often called "milk spots," are regarded as of inflammatory origin, although during life no symptoms of inflammatory action have been noticed. They are composed of fibres like those of the connective tissue, and acquire a firmness and hardness that is not observed in that tissue while in a healthy state. When old they may become calcareous.

The serous membranes throughout the body are nourished by vessels on their attached surfaces, branches of which penetrate the structures covered by them. It is not surprising, therefore, that those structures should partake of diseased action that is begun in the membrane. Hence, when pericarditis is acute and fatal it is not uncommon to find an œdema of the outer layer of the heart fibres, a paler color and softening. This will naturally diminish in some degree the force of the heart-contractions during life. This is a limited carditis secondary to the pericarditis.

The *symptoms* of pericarditis are very vague and obscure. It often cannot be made out without the physical signs, consequently little was known of it except its *post mortem* manifestations, till the rules of auscultation and percussion were systematized. There may be a sense of weight and oppression in the chest. There will be an increased frequency of pulse and respiration. There may be a painful sensation produced by pressure in an intercostal space over the heart, or pressing the liver upward by the finger under the ribs, at or to the left of the epigastrium, but of unprovoked or spontaneous pain there is none or next to none.

We turn, then, to the physical diagnosis, which is always, or almost always, complete. For this purpose it is desirable to consider, first, the cases in which the fibrinous exudation is the principal, or, perhaps, the only inflammatory product. These cases are known as *dry pericarditis*. The second class of cases comprises those in which, though there is, perhaps, the usual amount of fibrine produced, the prevailing exudation is serum. In both of these classes there is probably an initiatory friction sound due to the dry state of the pericardium during the inflammatory congestion, which precedes the exudation. Dr. Stokes, a great many years ago, reported such a sound, and on his authority I readily believe in its existence, although I have never heard it. And yet it seems to me strange that it should have escaped me; I have so often listened in acute rheumatism for the coming of the cardiac complication. In some of the many hundred cases of this kind, spread through half a century, it strikes me that I ought to have heard it; but I must say that I have not. When the serum or *liquor sanguinis* escapes from the vessels these will act as a lubricant and extinguish this rubbing till the fibrine is coagulated in the latter, then a new friction sound is developed by the fibrine. This friction is at first heard in the contraction of the heart. A few hours to a day later it becomes double, being heard in systole, and repeated in diastole. Commonly the first of these friction sounds is more distinct than the second, though both are easily recognized. This is, doubtless, due to the greater force of the systolic action, at first causing sound with a smaller quantity of fibrine than is recognized by the diastole, and afterwards showing the preponderance due to its strength.

The fact that the friction sound is first single and then double is of the first importance. Any physician, young or old, may, when he listens for the first time to a particular patient and finds an abnormal sound, doubt whether the sound is produced by a valvular change or external friction, or as Dr. Latham would have said, whether the sound is "exocardial" or "endocardial," and generally the tone of these sounds is not so characteristically distinct as to base an opinion on difference between them. But there is no endocardial murmur that can be single to-day and double to-morrow. This peculiarity belongs to the exocardial or friction sound. The changes in the form of the valves that produce murmurs are slow, gradual, and in acute endocarditis if a murmur is produced at all it is never double till weeks and sometimes months have elapsed from the time of the attack. The fact then that the abnormal sound was single yesterday, that is, one sound for one heart beat, and is double to-day, one sound in systole and another in diastole, is all but diagnostic of acute pericarditis.

But if the physician does not see the patient till the friction gives a double sound, how then? There are double endocardial murmurs occurring in the same times as the exocardial friction sounds. If the sounds are endocardial the history of the patient will be important; if he has rheumatism, has he had it previously? Has ascending or active exercise produced unusual shortness of breath? Is the heart enlarged? You may, even if there is fluid effusion in the pericardium, answer this last question by the aid of auscultatory percussion. In a word, has the patient had heart disease previous to this sickness? If he has not, the presumption is in favor of pericarditis. If he has chronic disease of valves and double murmur, the tone of the sound may be distinctive or it may not. The tones of old valvular disease vary very much; sometimes the sound resembles that produced by sawing; sometimes that of filing, or rasping. If the sounds, one or both are harsh they are not the sounds of friction; if they are of softer tone but harsher than the breathed "a-w-e," they may be produced either within or on the outside of the heart; if it is blowing, it is probably endocardial.

The friction sound is of a pretty uniform tone when

PERICARDITIS.   49

heard in different persons. It was called by him who first described it the noise of new leather, but it is not nearly so coarse or harsh as that. The sound caused by placing the palm of one hand over the ear and rubbing the back of that hand with the finger of the other is a nearer approach to it, but is only an approach, it is not the sound. The friction of pleurisy resembles it in tone, and is very apt to be produced in waves, the inspiration being attended by an intermittent sound made by two or three ceasings and renewals of sound. This is not often heard in pericarditis. If it is noticed its diagnostic value will be appreciated, as it does not occur in endocardial murmurs.

In 1850, or thereabouts, I was able to add a fact in aid of this distinction, which has since been announced annually in my lectures, but so far as I remember has not appeared in print. I should say, so far as I know, for many of my lectures have been reported for medical journals which I do not habitually read. The lecture is oftener taken without permission than with it, and after it is printed neither the reporter or the editor has the grace to send me a copy of the journal. I often learn the fact from physicians who make by letter or otherwise inquiries about ideas that I did not know had appeared in print. The fact to which I refer is this:

At the time already referred to I was examining a woman who had just been admitted to hospital. She had pneumonia and pericarditis. (I had by that time learned to search for pericarditis when I found pneumonia.) In listening to the heart sounds, it struck me that I heard the friction most distinctly on the turn of the respiration from inspiration to expiration. I listened till I was assured that this was a fact, then requested those who were with me to verify it, while I reflected on it. The thought was, why not? In inspiration the ribs are lifted up, and their pressure on the heart is at the minimum. The moment the inspiration ceases the rib begins to descend, and consequently make some pressure on the heart. This would make the contact of pericardium and heart more close. But if this is the cause of the louder sound, all the more will it be increased when the patient takes a deep breath and holds it while we listen. I immediately tried this method on this patient. It required only a minute to ascertain that each time the breath was held in this manner the sound was increased in quantity. We corroborated this observation day by day, so long as the friction sound remained in this patient.

The next person to whom I could apply this method was the son of one of my colleagues. He had pericarditis without pneumonia, and I was disappointed to find that when he filled his chest and then stopped breathing the friction sound was not only not increased but was actually diminished. What now becomes of my little discovery? Was it no discovery, or was there something in this patient in which his make-up differed from that of the first? I had him take a long breath and hold it while I percussed over the heart. There was an unusual amount of resonance. From this it appeared that the fully expanded lungs overlapped the heart, and that a half breath held in the same manner might give a different result. I tried it and got the same increase as in the first case. When the breath is held in any condition of inspiration or expiration, the pressure of the walls of the chest on its contents is increased.

Further study in this direction taught me that endocardial sounds are not increased by a full inspiration held, but if modified at all they are diminished.

These observations were made between thirty and forty years ago. The rules drawn from them have been announced to medical classes annually since that time.

PERICARDITIS.   51

They are, to me at least, the easiest and most rapid means of distinguishing pericarditis when there is a friction sound. But if there is none? In the greater number of cases the physician will find it at his first visit. After a day or two it will disappear, and after five to seven days more it can be heard again. On its reappearance it has but a short duration—one or two days. The explanation of these physical changes is very simple. Early in the inflammation there is exudation of fibrine and the pericardium is in contact with the heart. Hence the friction. By the third or fourth day the pericardium is lifted off the heart by serous exudation, and anteriorly, at least, there is no contact—no friction. After a time, five or six days, the inflammation has so far abated that the serum is in great degree absorbed and there is again more or less contact of the pericardium and heart. Both surfaces are rough and the friction is renewed with its proper sound. This renewal is heard first in systole, and then both systole and diastole. In other words, it becomes double in the same manner as the first friction sound does. But this return sound soon ceases in diastole first, and then in systole, because of the coalescence of the pericardium and heart. Thereafter there can be no more friction sound till these parts are separated one from the other in the manner already described. Regarding the first friction sound, I said just now that when the serous effusion lifts the pericardium off the heart the friction ceases, at least on the front of the chest. I had in mind at that moment a medical student who attended lectures for several days after he was attacked by pericarditis. He did not feel well, to be sure, but his chief complaint was of dyspnœa in mounting the stairs. I examined him while standing erect and while lying down. I found a large region of dulness on percussion, but no friction sound; but placing him on two chairs on his back, so that

I, sitting on the floor, could apply my ear to the inferior end of the scapula, I then heard the friction distinctly enough. Turning on the same two chairs so that I could from below upward place my ear on the cardiac region, I again heard it. In other words, the specific gravity of the heart is higher than that of the fluid which bathes it, and it sinks to the most dependent part of the sac. The flexible stethoscope is convenient for such an examination.

An acute serous exudation into the pericardium, in the absence of dropsy of other parts, is about as sure a sign of pericarditis as the friction. It is important, therefore, to know how to recognize it. A friction sound recognized and ceasing after two or three days can hardly be made silent by anything except serous, or purulent or seropurulent fluid. Here, then, is the first evidence of its presence in pericarditis. I have already given the dimensions of the heart and the points on the chest which correspond with its boundaries. Now, if the region of dulness exceeds these limits in any noticeable degree it is caused by hypertrophy of the heart, or by dilatation, or by morbid growth on the organ, or by fluid in the pericardium. If the disease is acute and the region of dulness grows day by day, it is fluid effusion. Ascertaining the extended dulness, you may, as I have already suggested, learn by auscultatory percussion how much of it is made by the heart and how much by something else. Observe in this specimen the relations of pericardium and the great vessels. The serous pericardium of the heart ascends on the anterior face of the aorta an inch or more before it is reflected off on the inner surface of the fibrous pericardium. In other words, the pericardial cavity extends upward above the base of the heart an inch or more. This extension upward, however, is confined to the anterior face of the organ, and the space given to it is only about one and a half inches wide. Yet this limited extension of the cavity, perhaps, I may say, may give you unexpected assistance in your search after serous effusion. The heart, as I have just told you, is heavier than the fluid of the pericardium, The fluid, therefore, will not displace the heart, but will rise above it and fill the portion of the cavity just described before it distends the pericardium laterally. The base of the heart is fixed at the line of the third rib, and bound as it is by vessels given off upward and by others coming into it from above and from below, no ordinary power will move it. For present purposes you may assume that as immovable. A dulness occurring above the third, extending to the second, will signify fluid effusion, and as it gradually but pretty rapidly extends the heart bag to the left and *to the right*, you cannot doubt that you have to do with pericarditis. But there are other means of diagnosis.

The enlarging heart sac will of course push the lung to the right and to the left of its natural position, and this will diminish the amount of respiratory sound heard over the proper pericardium, so that it will appear to be distant or be inaudible. Next, then, will be enlargement or fulness of the pericardium in proportion to the quantity of fluid effusion. In young persons this will be most evident. It is in them that the "pear-shaped" swelling is most often seen. This swelling has its neck upward, formed in that portion of the sac that extends above the third rib, and base downward on the diaphragm and to the left. In persons of mature age this "tumor" may be no more than a visible fulness of the front on the left side, a convexity rather than concavity of the intercostal tissues. This difference is referred to the yielding nature of the ribs in young persons and their firmer character as age advances.

The quantity of serous effusion that may occur in pericarditis is set down as *three pints and more;* when the quantity is very large another thing may occur. I was many years ago, requested to visit a patient half a day's journey up the river. I could not go on the day the application was made, and requested a young physician, who had lately gone through a hospital service with great credit and in whose skill and accuracy in the interpretation of the physical signs I had full confidence. When he returned he informed me that the lady had pneumonia of the left lung. I visited her the next day and found bronchial breathing and bronchial voice enough for the worst of pneumonias, on the left side behind, but no breath sound of any kind could be heard on the left front. The patient had had no chill, no rusty expectoration, and the dulness could be followed from the place in front where there was no respiratory sound to the line where the bronchial breathing became distinct; from that to the spine the dulness was marked, but less than up to that line. Then there was prominence of the pericardial space and dulness extending upward to second rib and under it. In short, it was a case of pericarditis with large pericardial fluid effusion, and not one of pneumonia. This gentleman was then skilful and has risen to great distinction. He could have made the diagnosis, but the signs of pneumonia so impressed him that he shut himself off from the other possibilities. A full inquiry into other conditions that could produce these signs would have set him right. I have said that the rational symptoms of pericarditis are in the main of little worth, but such a case as this is an exception. In this lady the distress for breath was pitiful; she could not sleep in the recumbent position; only in a chair. She could not bear the weight of this sac of water on her lung, in addition to the displacement caused by its bulk.

I have already narrated one case which shows how the bodily strength may be maintained while pericarditis is making its invasion. Another may be instructive. A young gentleman, a lawyer by profession, left his home in Vermont with the intention of transacting business in several cities When he reached Troy he felt ill; he could hardly say how, but his breathing and pulse were more frequent than usual. He sent for a physician, who told him that he had pericarditis, and bled him and also advised him to return to his home. This advice he did not accept, but the next morning went on his journey. I saw him in New York and found that the opinion given in Troy was correct and that he had already considerable fluid in the pericardium. I advised him not to go home, but to go to bed. This advice he would not follow, but persevered in the transaction of his business. He spent three or four days in New York, went to Philadelphia and Boston, and reached his home fourteen days after his disease was diagnosticated in Troy, all the time engaged in business.

There are a few cases of pericarditis that run their course so quietly, with so little local and general disturbance, that nothing is known of it till the evidences are revealed after death—death having been caused by some other disease. From this degree of mildness the symptoms vary all the way up to the distress and suffering observed in the lady whose case was just now narrated. Commonly you will find the patient lying on his back, his breath twenty-five or thirty in a minute; the pulse eighty to one hundred in the same time; and the temperation 100° or 101°—not a great sufferer, though Louis and Valleix found actual pain oftener than I have, that is, in about half of their cases. He keeps his bed, not because he cannot walk about, but because he is more comfortable in it. Bouillaud at one time attached importance to a noise in the heart beat which can be imitated by placing the palm of the hand over the ear and percussing the back of it by a quick blow from one finger of the other hand.

But when it was found that a heart made irritable by any cause gave the same sound, its diagnostic importance was greatly reduced.

The circumstances under which the rational symptoms are the least diagnostic are, probably, those observed when pericarditis is ingrafted on acute rheumatism. There is already fever and the common symptoms of it, and whether there is more or less of it, in these days of auscultation no experienced person allows himself to be guided by them alone; on the contrary, he pays little heed to them and forms his opinion from the physical signs. This is nearly or quite as true when pericarditis becomes a complication of pneumonia.

The mortality of pericarditis, uncomplicated, is small, almost nothing. Latham had no death from pericarditis alone in the cases already cited. Although he had three from pericarditis and endocarditis combined. When it occurs in the course of acute rheumatism, the two together are rarely fatal; but when it occurs in the œdema of scarlet fever, it is always grave and often fatal. The same is true of its coincidence with granular kidney. When it occurs with pneumonia it is difficult to estimate its importance, but it can hardly be doubted that it makes some cases of that disease fatal that without it might have recovered. Single pleurisy, not reaching empyema, is very rarely fatal, and it does not seem that coincident pericarditis adds to it a fatal weight.

When pericarditis is secondary, it is a question of some interest, at what stage of the primary affection is it developed? In the œdema of scarlet fever, if it occurs at all, it may be looked for in seven to ten days from the beginning of the swelling. It may come later, but the danger of this complication diminishes daily after ten days. With the granular kidney it has no fixed time. In a very large proportion of the cases of this disease it does not occur at any time, and when it occurs the event is associated with scanty and high colored urine and an imperfect separation of the urea from the blood. In other words, although it is not often found in those who die of the convulsions so frequent in the advanced stage of this affection, still the same condition of the kidneys seems to be associated with both, and this condition is often induced by imperfect protection from the cold, or the use of alcoholic liquors. Therefore, to know when there is danger of pericarditis in granular kidney we must know when the patient is going to be imprudent. It is not often in the first year, but when after two or three years the patient, finding himself still alive and comfortable, begins to doubt his danger and become impatient of advice.

In pneumonia it is only in exceptional cases that pericarditis occurs, and nobody has attempted to fix the day of its occurrence. I have only very rarely seen the concurrence in private practice. It is in the hospital that I have become acquainted with it chiefly. Persons attacked by pneumonia do not often get into the hospital before the third day of their disease. In no instance that I can recall was the pericarditis developed after admission, but in all cases it was found in the first examination of the patient. This is equivalent to saying, as far as a limited experience can say anything, that it belongs to the first three or four days of pneumonia, but it does not inform us whether the two diseases arise from one diathesis and are contemperaneous in their attack, or whether the pericarditis is secondary. You have been informed that albumen appears in the urine of all cases of severe pneumonia early, at least as soon as there is consolidation, and that in its course symptoms of uræmic poisoning are not uncommon. It is possible that this pericarditis is caused by this same poison. Whether it is or not is still a problem to be solved.

In the few cases of concurrent pleurisy and pericarditis writers are commonly satisfied with assuming that the inflammation extends from one membrane to the other, and they do not state from observation which is primary, but leave us the inference that the pleuritic inflammation extends to the pericardium. If this is really the fact, then we may turn to it for an explanation of the concurrence of pneumonia and pericarditis, for it is very rare that pneumonia is not a pleuro-pneumonia, and the *post mortem* evidences of pleurisy are often more marked than they are in simple pleurisy. But if the extension theory be correct, why is it so rarely illustrated? for there are hundreds of cases of either without any sign of inflammation in the other. Indeed, I think I may say they do not concur once in fifty, perhaps a hundred times. I do not remember that I have ever seen it, and therefore have no faith of my own to state.

Regarding rheumatic pericarditis, however, I am on very different ground; with this I am familiar. While it is true that the rheumatic diathesis is in rare instances declared by pericarditis before the pains appear in the joints, and these do actually follow after some days; the great rule is that the articular rheumatism precedes the cardiac affection. Such an exception was reported by my predecessor in this chair. I have witnessed two or three cases, and a case is recorded now and then by writers. These cases, then, are so rare that they deserve little more attention than the announcement of the possibility; especially as there is nothing in them to inform us of their rheumatic origin, nothing to suggest a treatment of rheumatism, any serous membrane may become inflamed in the course of rheumatism, but the pericardial serous membrane is attacked ten or twenty times more frequently than any other. This is, perhaps, accounted for by the fact that articular-rheumatism is an inflammation of the fibrous tissues of the joints, and that this membrane is the only one that is in contact with and attached to a pretty firm fibrous structure—the fibrous pericardium. This is not identical in tissue with that of ligaments of the joints, but there is a resemblance between them nearer than between any other two tissues of the body. This suggests the idea that the fibrous pericardium may attract rheumatism as fibrous structures of joints, but less strongly, and transmit it directly to its serous lining. The time when this complication occurs is not commonly till the fifth day after the joints are attacked. From this to the eleventh day the liability is greatest. This, in my view, is an important fact. For if rheumatic pericarditis delays till the fifth day of the articular disease, and if we can cure the latter in less than five days, can we not save hundreds of hearts? Since the salicylate of soda has been relied on for cure of rheumatism, many cases are cured in one day and few persist till the fifth. That is, the inflammation is subdued, but stiffness remains for a day or two more. I think I have seen many instances of this protection of the heart. Of European writers some admit this; others deny it. To my mind, this difference arises from the fact that they do not all use the drug in the same way, and in the same doses.

Treatment of Pericarditis.—As late as 1840 the treatment of pericarditis without a mercurial was generally regarded as hardly less than malpractice, and it was to be carried to the production of redness of the gums and slight salivation. Dr. Latham, whose pleasing style won for his book a world of readers, knew no other way, and urged his views with great earnestness. The idea was that the mercurial "diminished the plasticity of the blood" and prevented the plastic effusion, and thus gave the patient a better chance of recovery. Soon after the publication of this pleasing book Dr. Taylor published the results of a different treatment in another hospital. He gave no mercurial whatever, and his results in, it may be, fifty cases were as good and, as I remember, better than those of Dr. Latham. This greatly surprised the advocates of that treatment, and was the beginning of its abandonment, as a plan. Now a physician, who has charge of a case of pericarditis hardly thinks of a mercurial of any kind, and it is not given in one case out of twenty.

The first consideration, as the disease is so often secondary, is the treatment proper for the primary affection. If, for example, it follows articular rheumatism, the question will be, what will most quickly and safely eradicate the rheumatic diathesis? This leads to some statements regarding the treatment of articular rheumatism. Until Dr. Fuller's alkaline treatment was announced we were much at sea; hardly two physicians agreed on any plan. One relied on nitre, in large doses, and greatly diluted. Another thought that lemon-juice, given freely, gave him better results. Another gave quinine. A fourth relied chiefly on diaphoretics, and still another had faith in the external application of boiled boughs used as hot as could be borne. But treated by whatever plan it was but slowly cured, if cured at all. It often lingered for two and three months. But the adoption of Dr. Fuller's plan was attended by a notable shortening in the duration of the disease, which was cured in three to six days in a large proportion of the cases. A few resisted, and the disease was not controlled for weeks. The plan had memorable advantages over any previously proposed, and was in detail this, or rather is this: the administration of an alkaline medicine in full doses so as to make the urine alkaline in the shortest time consistent with the toleration of the stomach. He used sometimes the Rochelle salt, tartrate of potash and soda; sometimes the

PERICARDITIS. 61

carbonate or bicarbonate of soda; sometimes the same salts of potash. He had the urine tested at short intervals to ascertain when this saturation had occurred. After such saturation the urine was to be kept alkaline till the symptoms abated, requiring, usually, smaller doses than were needed to produce the alkalinity. This is the pith of the matter.

About ten years ago the salicylic acid was introduced for the cure of rheumatism, and the physicians of Bellevue Hospital, prominently Dr. Jacobi, desired to try it; but how to dissolve it was the question. He proposed the carbonate or bicarbonate of soda as the solvent. The acid was mixed with water first, and then the soda was added till the mixture became clear. This was the first salicylate of soda made in this country, so far as I know, and with it the disease was treated with marked success. I published a dozen cases, or so, soon after this preparation came into use, showing the almost marvellous promptness with which this medicine gave relief. I can narrate some others, indeed many, but will not be wearisome.

A man, of about middle age, was brought into one of my wards when I was present. I found his right knee swollen, red, hot, and painful, so that he could not move it. The disease had been upon him two days. He had some fever. Salicylate of soda was prescribed, twenty grains in solution every two hours. No other medicine was given. The next day at the same hour the swelling, redness, and pain were gone; only stiffness remained. The medicine was continued, but diminished to half the quantity. The next day I found him sitting up and he could walk. On the third day he asked for his discharge, and was discharged. I heard no more of him.

A young lady of remarkably handsome figure and face, before I knew her, had had three several attacks of rheumatism from one to two years apart. She had been confined to her bed, or arm-chair, by each invasion, an average of three months, and in each the heart had suffered. She came with her mother to the city for a stay of a few days. Her mother had previously, and again at this visit to the city, consulted me regarding her own ailments. One morning she came to me with a very sad face, and told me that her daughter, since the day before, had been brought to bed with a new attack of rheumatism, which appeared to be the worst she had ever had. Both knees, both elbows, and one ankle were red, painful, and swollen, and that she was utterly helpless, "and now," she said, "my poor child must suffer another three months of agony," and she wept. I bade her be comforted, told her that the world moves and men make discoveries, that "the thing that hath been, is" not now, always, "the thing that shall be," and that by the use of a new medicine, I hoped to give her daughter pretty prompt relief. It was ten o'clock in the morning. I prescribed the salicylicate of soda, fifteen grains every two hours. I could not visit her till nine o'clock at night. I was happy to find her almost wholly relieved of pain, and the mother informed me that the swelling and redness were diminishing. She rested well that night. The next day at 1 P.M., I found her sitting in a chair at the side of her bed free from all the symptoms of rheumatism except stiffness in the invaded joints. The third day she could walk. Thus ended her rheumatism, and thus was lifted from her mother's heart a load of sorrow. That was the end. I saw her three years later. Her old heart disease remained, but there had been no more rheumatism.

These are selected cases, and selected for the purpose of showing with what energy and promptness the medicine acts in certain cases; and especially to show that the drug can, in many cases, at least, strangulate a rheumatism before the average period arrives for the occurrence of heart complications.

There exists in my mind no doubt that the promptest, best, and safest remedy for articular rheumatism, and consequently the best prevention of its secondary cardiac disease, is the salicylate of soda. Still there are certain facts which should be stated *per contra*. Given in the quantity just reported it does, in some persons, either by its direct action on the stomach, or through the nervous system, produce vomiting; then it is given in smaller doses, or it is suspended for a few hours, and the bicarbonate of soda given instead.

Early in the use of the salicylate of soda, it was noticed that now and then delirium followed its administration, and seemed to be produced by it. This effect seemed to be chiefly observed in persons who had abused themselves by too free a use of intoxicating liquors. This delirium occurred almost exclusively in hospital patients, whose "previous history" it is often impossible to get. The delirium is violent, yet it was, commonly thought prudent to send these patients to the wards where the delirious are treated. This mental disorder lasted but a day or two, and has been, in no case, followed by any permanent disease. In these, again, the carbonate or bicarbonate was given instead of the salicylate.

The salicylic acid has been given alone. It will cure articular rheumatism, but less promptly, it appears to me, than the salt. The European reputation of the acid led Dr. Jacobi, almost without knowing it, to make and use the salicylate, and, for myself, I must say that after several years use of it, it has, at present, my decided preference.

The abstraction of blood is hardly ever thought of in the treatment of articular rheumatism, so it is almost never resorted to for the cure of rheumatic pericariditis

The aim is to overcome the rheumatic diathesis and so cure both.

When pericarditis occurs in the œdema after scarlet fever, it is probably produced by uræmia, or the same condition of the system that causes the dropsy, and is almost always attended by a high-colored, or bloody, or smoky and scanty urine. The first object of treatment, then, is to remove the cause of the disease; in other words, to induce a healthy secretion of urine, if that is possible, and remove from the system the poison that is being carried into every part of it, whether that poison be urea, carbonate of ammonia, or whatever else. To accomplish this the most valuable agents are the diuretics and diaphoretics. Among the first, I give preference to an extemporaneous citrate of potash and digitalis. You prescribe half an ounce of the carbonate of potass, dissolved in six ounces of water, and direct that a tablespoonful, mixed with the same quantity of fresh lemon-juice, be given every two hours, to an adult, with the proper reduction for children. On the mixture of these two ingredients there is a feeble effervescence, but enough to make the draught agreeable to the taste. It is not a new prescription. The physicians who were old when I was young gave it greatly more than does the present generation. With this give an adult a dessert-spoonful of the *infusion* of digitalis three times a day. I emphasize the word infusion, because it seems to me to act much better, as a diruetic, than the tincture, or extract, or powder, and because many physicians have thanked me for the suggestion. The quantity of the infusion may be larger than a dessert-spoonful. The authorized dose is a tablespoonful, but the smaller portion has in most cases been enough; and then, the only cases in which I have seen what are called the "cumulative effects" of digitalis were in uræmic persons, and I am afraid to give it freely to such persons. These two medicines act better when given in this way than when either is given separately. The potash salt, probably, causes an increase of water of the urine, and the digitalis the solid, or solidifiable elements.

Everybody knows the relations that exist between the skin and kidneys, and you will, therefore, be prepared to hear that when those organs are congested a certain treatment of the skin may help to bring relief. What I am about to advise is more preventive than curative and is, therefore, all the more valuable. As soon as the œdema and paleness appear, use the warm water or vapor bath until a perspiration is produced. At first the perspiration may be a little free. In this condition put the patient in bed without losing much time in wiping off the water. Keep up the sweating for half an hour, then let it dry down to a mere moisture, and keep the patient in this moisture for days, even for two or three weeks, by the bed-clothes. In many cases of children it is almost impossible to keep the patient in bed. He does not feel sick and cannot understand why he should not be allowed to run about as usual. In such cases I have adopted the plan of getting the first perspiration and keeping the child in bed for a few hours. Meantime a double suit of flannel is prepared—pretty heavy flannel—and when he will not tolerate the bed any longer he is clothed in this suit and allowed to walk about and enjoy his playthings; but with this one important condition, that the temperature of the room or rooms into which he is allowed to go be kept at or above 76° F. This temperature, this clothing, and the exercise he will get will be likely to keep a little sweat on the forehead and body the whole time. This is what you will try to produce and keep up for days together. You may expect from this management a diminution of the congestion of the kidneys, and with that a freer secretion of urine, and a gradual diminution in the morbid indications of that fluid, and by this the danger of convulsions and *pericarditis* is naturally reduced. If pericarditis has actually begun, the patient will probably remain in bed willingly, and the perspiration may be induced by the footbath in bed, and kept up by repetition. The vessel holding the warm water is placed in bed close to the body, the thighs and knees so bent that the feet will fall perpendicularly into the water, and all covered by the bed-clothes. This is the first step in the treatment, and may be followed by the depurative diuretics. The condition of your patient does not suggest the lancet, but rather tonics, including iron, and if you take any blood at all, it should be limited to that which a few leeches will abstract. There is probably no therapeutical fact more fully demonstrated than the efficacy of *blisters* in pleurisy. Are they equally useful in pericarditis? I do not think so. In one case the blister is applied near, physiologically, as well as in actual distance, to the diseased surface. In the other the physiological distance is much greater. The vessels that supply the heart with blood have their origin from the aorta, before it leaves the heart sac, and the dependence of the sac itself on the thoracic arteries is not intimate. A blister applied over the heart may act as a derivative on other branches of the thoracic vessels, but probably not at all on the coronary arteries of the heart. However this may be, blisters are not a favorite application in this form of pericarditis, perhaps because it occurs so generally in children, and the student is admonished to use them with modification and great care in these young folks, on account of the exaggerated and even dangerous effects they may produce.

Opium deserves consideration in this form of the disease. It may not, by any means, be given in heroic doses, as it is given in the treatment of peritonitis, for a reason which I will soon explain to you; but given in ordinary doses, or only slightly more, it soothes irritation, and the heart is always irritable during the early stage of pericarditis, and gives the patient needed rest, and in this way gives earlier development to that recuperation and recovery which in many cases may be regarded as the spontaneous tendency of the disease itself. The objection to large doses of any opiate in the course of uræmia in any of its manifestations may be enforced by the following sketches. The late Dr. Conant Foster, living near me, sent for me early one morning to see a young man who had slept in his house. He was comatose, but not profoundly. The history in short was this: He sought the doctor's advice the night before on account of a great pain in one of his fingers, which was the seat of a felon. The doctor advised him to take forty drops of laudanum on going to bed. In the course of the night the patient sent a message to the doctor, informing him that the pain was very severe and he could not sleep. The doctor advised him to take another dose of laudanum of forty drops. In the morning he was found in the state already indicated. He could not be aroused to answer questions, yet he was not motionless in bed; the pupils were contracted, but not extremely; he did not swallow except unconsciously; the breathing was not remarkably deep or infrequent; it was not the characteristic stupor of opium-poisoning, but more like that of uræmia. Dr. Foster did not know of any illness before the felon occurred. There was no œdema. The chance that any laudanum could be found in the stomach was hardly worth entertaining, since the last dose was taken five or six hours before, and the stomach was empty when it was taken. The organ was washed out by the pump, and no odor of laudanum was discovered in the water used. The remaining indication was to procure free perspiration. I have forgotten whether this was done by enveloping the patient in several thicknesses of blankets wrung out of hot water, or by use of the foot-bath in bed; but our efforts were unavailing. He died that day. A post-mortem examination revealed what he and his friends appeared to be wholly ignorant of, a rather advanced stage of Bright's kidney.

During the week in which I saw this patient I was called to see a lady, in consultation, who showed evidences of Bright's disease in œdema and paleness, manifestly. She had the night before taken, at one dose, half a grain of sulphate of morphine, I do not remember for what. She was in a coma similar to that in which the young man was found, not having the close contracted pupils, the deep slow breathing, the motionlessness of the opium poisoning, but again more like that produced by uræmia or the early stage of miasmatic coma. I have seen other similar cases, enough to persuade me that there are conditions, not readily recognized, in uræmia, in which the blood poison and opium may work together, and produce a coma which neither of them alone would have produced, at least, at the time when it occurred. On the other hand, many patients having Bright's disease have taken, under my eye, but not with my approval, half a grain of the sulphate of morphine, night after night, with only soothing and sleep-giving effects. This was the case with my colleague, the late Prof. Gilman. He gave me all the confidence I could ask, except in this one particular. I assented to one quarter of a grain of the sulphate at night, but he would take, night after night, a half grain, to which dose I gave my sanction, only after the proof that for him at that time it was not dangerous.

These cases, it is true, are neither of them pericarditis, but all Bright's disease, and if pericarditis occurs with that affection, either acute or chronic, they illustrate a needed caution in its management.

The co-existence of pericarditis and pneumonia, or of pericarditis and pleurisy, is of infrequent occurrence—so infrequent that it does not suggest the thought that one is secondary to the other, as we are apt to say pericarditis is to rheumatism. It seems to be a concurrence, as if the same diathesis had produced them both. There would seem to be then nothing distinctive in the treatment of such a pericarditis. What we would do for one we would do for the other. Then for the treatment of pneumonia. When I entered the profession bleeding from the arm was almost invariable. The patient was bled as soon as the diagnosis was made. Commonly sixteen ounces of blood were taken, or the blood was allowed to flow till the patient sitting on his bed grew pale, and the sweat bedewed his forehead. In twelve hours the same thing was repeated, and after another interval of twelve hours, ten or twelve ounces were taken by cups or leeches. Bleeding, more or less, in pneumonia was so general that its omission was regarded as hardly less than culpable, and the first implement with which the student supplied himself was a lancet. In 1837 I followed Bouillaud in Charity Hospital, Paris, and witnessed his *coup sur coup* bleedings in this disease. It consisted in bleeding from the arm to sixteen ounces in the morning, taking twelve ounces by cup at noon, and sixteen ounces by venesection at night. This practice was repeated for two or three days, so that to abstract eighty to one hundred and twenty ounces of blood in the course of three days was common with him. He was not indiscriminate in his application of this treatment, for if the patient was broken down, or was of bad habits previous to the pneumonia, he was not bled at all, but was put on the tartar emetic treatment so much extolled by Laennec. It cannot be denied that many of his patients lived through it, possibly were cured by it. But it was not just to contrast the results of these two modes of treatment, after the care he exercised in the selection of his cases. Physicians generally were not converted to Bouillaud's plan, though at that time they were all bleeders. They seemed to think that it was "too much of a good thing."

Eighteen years, or thereabouts, after this, good old Dr. North, who had retired from active practice, spent a winter in New York, and on clinical days was always at the hospital. By that time the sanguinary habits of the profession had been materially modified. Venesection was practised with discrimination. I had almost abandoned it in hospital practice, making, however, frequent use of scarification and cups. The good man, after hearing the directions for the treatment, would fall back and, getting the ear of one of the class, would say, "This will not do. We cannot cure these men without bleeding from the arm." At the end of the season I was able to inform him that he had seen in my wards forty-three cases of pneumonia, out of which there were but three deaths. So small a mortality I then considered "a run of luck," in the character of the cases, and have not changed that opinion, for if it was skill in treatment, that skill departed from me after that winter.

The late Dr. A. H. Stevens had convinced himself that the abstinence from venesection by the whole profession, with the exception of here and there "an old-fashioned man," was made necessary by a change in the human constitution, and that in his later years man could not bear the bleedings that he witnessed and practised in earlier life. · The practice of the few remaining bleeders; the hemorrhages of accidents and of parturition, with subsequent recovery, were cited to him; but he would reply, "You do not know about it. I have lived through it all, and know it all well."

PERICARDITIS. 71

Subsequent to this conversation a very striking case occurred, which, had he witnessed it, might have shaken his conviction. A very eminent physician, when turned of seventy, was attacked by pneumonia. The lung—it was the right—was in the crepitant stage when I first saw him in consultation with late Prof. J. M. Smith. We had a dozen leeches applied in the morning, and met again in the evening. The first thing that struck us on entering the room was the extraordinary paleness. There was little difference between the color of his face and that of the pillow on which his head rested. His expression was languid. It instantly occurred to us that the leech bites had been forgotten by the attendant and that the patient was bleeding to death. I at once examined and found it so. Our first care was, of course, to stop the bleeding. I pressed all the fingers I possessed on as many leech bites. The bleeding was stopped, say, two minutes. Then, turning attention to the amount of blood lost, his clothes, including a flannel undershirt on the back, were saturated. The blood had spread widely on the under sheet. From the ticking of the bed I removed three double handfuls of coagulations. It had saturated two mattresses for a pretty large space, and there was a large pool of blood on the carpet under the bed. It was impossible to ascertain the quantity he had lost by any measurement. But that it was enormous is shown by what follows. We drew the sheet out from under him, and after washing the back in warm water drew under him an india-rubber and a linen sheet, having previously torn the personal clothing up the back and laid it forward. We did not dare to allow him to sit up in bed. We were about to tear the fresh clothing up the back, put his arms into it, and tuck it under the back. At that point he said, " Give the undershirt to me. I am strong enough." One of the attendants gave him the shirt while our attention was diverted.

He partly raised himself in bed, and had partly passed one arm into a sleeve when he fell back fainting. When he recovered his consciousness he willingly submitted to our plan. His pulse could scarcely be felt, but when countable was 140 in the minute. For days, indeed for two or three weeks, even when he could sit up, he was the palest man I think I ever saw. This bleeding completely strangulated the pneumonia. The crepitation ceased after a few hours, and there was no hepatization at all. There were no remains of pneumonia the next day, and no more pneumonic expectoration, but he did not recover from the bleeding for some weeks. This was, by accident, heroic treatment, and, so far as the inflammation was concerned, was attended by unprecedented success. But I have not dared to follow its suggestions.

Among the most pleasing books in medical literature is "Bartlett on the Certainty of Medicine." In that there is a review of the treatment of pneumonia by bleeding, and the author proves, as far as anything can be proved in therapeutics, that the practice diminishes the severity and the mortality of the disease.

In the hospital I have rarely resorted to venesection for two reasons : first, the patients are most of them persons whose health has been impaired by the unwise use of intoxicating liquors or scanty or improper food. Second, they are likely to come in at a stage of the disease in which active treatment is harmful. On one occasion a young woman of good habits and of previous good health was admitted, on the third day of pneumonia. I asked the house physician if he had ever bled a person. He had not. Would you like to bleed this patient? He would very much. Bare up the arm and I will guide you. But he had no lancet. Other members of the house staff were present, but none of them had a lancet. There were several persons at the clinic—students and young physicians

—not one of whom had a lancet. Not one could be found in the hospital except among the surgical instruments. I mention this occurrence to show what kind of patients are benefited by venesection, and how even then, fifteen or twenty years ago, bleeding had gone out of use and almost out of memory.

But in hospital practice I still make constant use of cups with scarifications. My private practice has always been limited to consultations. I have never had a single family. The consequence has been that I have rarely seen pneumonia, or, for that matter, any other disease till the dangerous period had arrived, or for pneumonia till the stage for active treatment had passed. Still my faith in bleeding on the first to the fourth day is as active as ever.

When the period for active treatment is passed, that is, when the consolidation of the lung has extended as far as it will, which, in the majority of cases, is at the end of the fifth day, it is plain that the mischief is done. Bleeding is of no use so far as the consolidated lung is concerned, but even then it may be called for in aid of the other lung, for that other lung is in danger of congestion from the diversion of the circulation to it, caused by obstruction to the flow of blood through the vessels of the inflamed lung, and consequent œdema. A moderate bleeding may help to relieve this congestion and œdema. Aside from this we do not entertain the question of bleeding, in any way, after the end of the fifth day, counting from the initiatory chill. But while the disease is passing from the red hepatization to the gray, which every pneumonia must do if the patient is to get well, there are still active symptoms. A high temperature may require sedative doses of quinine, or aconite, or possibly veratrum. In many cases of pneumonia a condition of the system occurs which suggests the use of alcoholic stimulants. I doubt whether they ever do any good—whether a case that cannot recover without them will ever be saved by giving them. At the same time I do not think they do harm when properly used. What you would do for pneumonia you would do for pneumonia and pericarditis combined. For this pericarditis there is not a separate treatment unless it consist in giving moderate doses of an opiate to quiet the irritability of the heart ; and this may be done without fear of any unfavorable effect on the lungs.

When pleurisy and pericarditis concur, blisters and diuretics, which are so effectual in pleurisy, are not so useful, as has been said already, against the pericarditis. In addition to these the abstraction of blood by cups or leeches may hasten the cure of the pericarditis, especially as the opiates, so useful in most forms of that disease, may interfere with the action of the diuretics.

Pyæmic pericarditis has a peculiar history. It appears that a common preliminary is the formation of small pus spots in the muscles of the heart. That those near the surface provoke the pericarditis, and that the exudations are there lymph and serum as in the more common form of the disease. But it sometimes happens that one or more of these spots grow to a considerable size and rupture into the pericardial sac. The inflammation that follows this accident is likely to be attended by purulent exudation. There is little to be said of treatment in this variety of the disease. The primary disease is so apt to be fatal, and the oppression and weakness imposed on the system before the pericarditis manifests itself is much opposed to local depletion, and may forbid it. The soothing influences of opium suggest it, and it may be used, but not in heroic doses. When the effusion is purulent, should the patient survive the pyæmia, as is quite possible, the question of tapping will present itself.

Pericarditis may occur independently. It is rare, but a case showing its possibility will soon be narrated.

PERICARDITIS.   75

When it occurs in that manner the moderately free use of cups with scarification, or leeches or both, and of an opiate will give the best results. It is common to apply these to the pericardial region, but as it is not derivation that is now aimed at, but a moderate reduction of the volume of the blood, it would seem to be of little importance from what part of the body the blood is taken. Cups or leeches applied over the heart do not prevent the physical examinations which should be duly made, as blisters do. But blisters I am not disposed to urge. I have here a specimen which I believe I have not yet called your attention to, and while there is not a word to be said regarding the treatment of such a case, could it be recognized during life the curious fact which is here demonstrated is worthy of your attention. You observe first that there is a sensible enlargement of the whole heart. Second, that the pericardium is closely attached to the organ everywhere; and third, that there is a hoop of calcareous matter, firm as bone, completely investing the heart in such a way that you would suppose the heart could not move either in systole or diastole. You observe it here over the right auricle an inch in width and an eighth of an inch in thickness; it runs downward and to the left across the right ventricle on to the left, about an inch from the apex, passing up the posterior of the heart in a line nearly parallel with that of its descent, till it becomes continuous with itself over the right auricle. Here, as it turns around the left ventricle, it has been broken to give admission to the interior of the organ, and here it is formed into shape of a rope and is of the size of the little finger. You notice that the structure is not homogeneous as you would expect that of bone to be, but that the fracture is irregular, and parts are of a whitish color and other parts are yellowish. When it is not broken you do not see it because it is wholly within the pericardium, but you hear its bony ring when I percuss it, and you can feel it and follow through its whole circle as distinctly as you can feel a coin in your pocket. The pericardium is immovably adherent to it throughout, and the heart by a loose, thready tissue.

The specimen came into my possession without a history, and all we can learn of the ante-mortem facts must be obtained from the specimen itself, and these are scarcely enough. The evidences of a pericarditis are found in this adhesion of the pericardium to the heart, and that this occurred years before death is rendered probable by the thready state of the adventitious membrane, and still more by the presence of this strange hoop that was formed in all probability after and in the lymphy effusion. It is probable also that the ring was completed before it had acquired its present breadth and thickness. In other words, that it required years to get its present dimensions. Meantime the heart was beating. With how much efficiency we can never know.

I have never met another that at all resembles this specimen, and do not remember that I have seen a record of a similar case. But I have not made *extensive* search for it, so that I cannot say it is without a parallel in morbid anatomy, but it is safe to say that it is a very rare occurrence.

And while we are in the line of rarity I will give the case alluded to a little while ago.

An Irish laborer, aged 30, a well-developed man, was admitted into the New York Hospital on 16th of January, 1836. He appeared to have had an inflammatory affection of the chest tissues three weeks before. The heart region was edematous; patient pale and feeble, but not emaciated; pulse 130 to 140 in the minute and on exertion fluttering; has cough attended by slight expectoration of mucus; respiration oppressed; incapable of any exertion—even turning in bed increases his dyspnœa and causes the heart to flutter; tongue but slightly furred; bowels constipated; has nausea and vomits all ingesta; tenderness of epigastrium.

Marked dulness in pericardial region and to the right and left of it. When the patient turns on his right side the dulness extends an inch farther to the right and recedes from the left slightly. When he turns on his left side the dulness extends equally to the left. Not the slightest sound of respiration can be heard over this whole region, though distinct in other parts with an occasional mucous râle. The impulse of the heart, though not forcible, can be felt over all this region of dulness in its calmer action. Then the sounds are normal and its rhythm perfect, most distinctly heard at the base. The apex does not strike against the walls of the chest in any position of the body, not even when prone. A slight undulatory motion is perceived in the epigastrium and in the anterior parts of all the lower intercostal spaces at each pulsation of the heart. The action of the ribs in breathing is much restrained.

There may have been a peritonitis, but the man's present suffering arises from an enormous distention of the pericardium by fluid, and the heart's action is crippled by its pressure when exercise requires stronger beats.

The nausea and vomiting were supposed to arise from the condition of the bowels.

He improved on the use of a purge, and the nausea and vomiting ceased. Otherwise the treatment was by diuretics.

This patient remained in the hospital fifty-one days, being much of the time able to sit up, and sometimes to walk a little. On the forty-ninth day he fainted while at stool, and for two hours the face and neck were livid; pulse just perceptible at the wrist, but too weak to be counted; feet and hands cold. The next but little improvement; pulse could be counted only now and then, for the most part a flutter; face livid; respiration labored, slow, and deep; the mind clear. On the fifty-first day he was found to be dead one or two minutes after his pillow had been adjusted. The disease had lasted, out of the hospital and in it, seventy-two days. A part of the time there had been œdema of the feet and ankles. The region of dulness was enlarged in the last few days of his life.

Autopsy three hours after death:

Face and neck congested and livid; surface over the abdomen, thorax, face and neck œdematous; those of legs and feet were not; superficial veins bled freely when cut; some clear yellow serum in abdominal cavity; undigested food in the stomach, apparently all he had taken in the last two or three days.

The liver was forced downward so that its upper surface was where its lower edge usually is, at the free border of the ribs. The diaphragm was forced downward so as to make a very large convexity downward into the abdomen.

The distended pericardium was loose and hard, occupying the whole anterior of the chest and extending out of view on each side when the sternum and cartilages were removed. It extended backward to the spinal column and downward so as to form the pouch in the diaphragm and depress the liver as stated above. Each lung was pressed into the posterior and upper portion of the thorax. The edges of the lungs were adherent to the pericardium on each side. A small quantity of fluid was found in each pleuritic cavity. The lungs were healthy but compressed.

The pericardium when opened was found to be one eighth of an inch in thickness and to have a firm, leathery feel, and contained a gallon of clear yellow serum. There was a thick covering of lymph over all the heart and all the inner surface of the pericardium. The two were united by tendinous threads and bound posteriorly and anteriorly. There had been adhesion of the pericardium to the heart anteriorly, as was evident from patches of lymph found on one surface, removed from one surface and found on the other, and from flakes of lymph attached by one edge of the floating fin and by broken threads so attached. The deposits on the anterior of the heart were stained with blood effused underneath them.

This case is copied from the New York Hospital Record Book, and was then written by myself soon after the date given above. There are two other facts relating to the case which are not there recorded, both of which are as distinct in my mind as if the case had occurred yesterday. How the first escaped record I cannot explain, except on the supposition of haste, as house physicians then as now were crowded with ward and other work and had little time for " writing up cases." The second was purposely omitted, because it might have been regarded as an improper criticism by an inferior on the judgment of his superior.

The first of these facts was that in the last week of the patient's life, after the dulness was found to be extending, the sternum was found broken up into several pieces, which were movable by pressure of the finger, one on another. I cannot say that it was a disunion of its natural subjoints. It seems to me that the fragments were of irregular shape, but of the breaking up of the sternum I have no doubt. The second fact is that I asked the two attending physicians in succession for permission to tap the pericardium. They both declined to perform the operation, or to permit me to perform it, and when at length the patient had nearly reached the "in extremis" I applied to the superintendent for his sanction of the operation, but he declined to give it because he was not a physician and had no control in the medical administration of the hospital, and so the patient died, and probably would have died if the operation had been performed at that late period of his disease. This would not have been strictly a new operation, as will be seen hereafter, but was then wholly unknown to the hospital physicians and probably to the whole profession in this country.

Tapping the Pericardium.—In vastly the greater number of cases in which inflammatory action has caused fluid effusion into the pericardial sac, the question of tapping does not arise. Every physician has been surprised in observing how quickly, after the inflammation has begun to subside, the fluid disappears. Such cases take care of themselves, and there is but little respiratory oppression, and what there is is of short duration. In such cases the question of tapping is not entertained. When the fluid, however, compresses the heart and obstructs the entrance of blood by the venæ cavæ, and encroaches on the breathing space by its bulk, when the patient's life is in danger by the quantity of the fluid, then I must urge as I did in the case just narrated, that tapping is positively demanded. But is there not danger of wounding the heart? Undoubtedly, and it was for this reason the earlier operators opened the pericardium with bistoury and scissors, instead of a trocar. But regarding these wounds, the experience of New York surgeons in the last thirty years has in some degree diminished the dread of them. One man was shot through the heart and lived several days. After death the bullet was found in one of its cavities. Another man lived eighteen or twenty days after a bullet had been lodged in the septum ventriculorum. Of late the right auricle has been purposely penetrated by a hollow needle for the depletion of this cavity, and it is said that the patient was temporarily improved by the operation. Still we all feel that no fool's play can be permitted about the heart, and it is the aim of every operator to enter the pericardium, and at the same time avoid the heart. In almost all the cases requiring tapping, when the patient is in the prone position, the heart falls backward to the posterior limit of the sac, and the fluid comes in anteriorly between the sac and the heart. The depth of the fluid in this position varies from half an inch to an inch or more. It is possible to penetrate the pericardium anteriorly, and yet stop short of the heart itself. Otherwise, choose a point which the distended pericardium occupies, but which is off the limits of the heart.

Dr. John B. Roberts, in 1880, published a volume entitled Paracentesis of the Pericardium, in which he gives a table of all the cases in which this operation had been performed at that time, so far as he could rely on the printed reports. He finds in all sixty, in these the point of insertion was the third intercostal space in 2; the fourth in 22; the fifth in 18; the sixth in 5, making 47 in which the point of insertion is stated. Those made in the third and fourth spaces were over the heart; those in the fifth would have a good chance of escaping the heart, especially if made against the upper border of the sixth rib; and those in the sixth would be nearly certain of being off it, unless the organ was enlarged. In these sixty cases there were twenty-four recoveries. These are distributed as follows: Of the two in the third space, 0; of the twenty-two in the fourth, 10; of the eighteen in the fifth, 9; of the five in the sixth, 2. So three of the recoveries belong to the cases in which opening was performed by bistoury and scissors, or those in which the place of tapping is not stated. Success is about equally divided between the fourth and fifth spaces. With the fifth the recoveries were just one half, for the fourth a fraction less, or $\frac{5}{11}$. Twenty-four recoveries in sixty operations does not seem to be a brilliant success, but if we can suppose that twenty-four persons by this means were really snatched from death, it assumes another aspect. I have not attempted to look up and read up the cases which Dr. Roberts has tabulated, but if many of them were rescued from a condition anything like that which is detailed of the case in the New York Hospital, where I implored permission to tap the patient, the operation has accomplished little less than resurrection.

Dr. William Pepper's case (*Am. Jour. of Med. Science*, April, 1879,) resembled mine in severity, with this difference, that mine was *moribund* and left to die, his "was evidently moribund," was tapped and recovered, so far as the cardiac effusion was concerned, and died fifteen months later of an unusual form of disease of the serous membranes. The same diathesis which produced the latter may have and probably did produce the pericarditis, but at the inspection the pericardial cavity was wholly obliterated by adhesions, and appeared to have nothing to do with the fatal issue. He began with a most discouraging case, but he brought it out triumphantly.

His trocar was inserted "in the fifth intercostal space about one inch inside of the line of the left nipple, i.e., nearly in the normal position of the apex beat, and over eight ounces of reddish serum were removed." Relief was immediate, recovery was slow and incomplete, as she was weighed down by other disease.

As to the mode of operation, whether by what is now called "aspiration" or by the old-fashioned trocar is not very emphatically taught by Dr. Roberts' table. Of his whole, sixty, thirty-six died. Among these sixty, "aspiration" was practised in twenty-four. Of these, ten recovered and fourteen died; or, in figures, $\frac{2}{3}$ of the whole died, and $\frac{7}{12}$ of those who were "aspirated." "Aspiration" of the pericardium does not appear to have been practised till about fifteen years ago. Since then it has been preferred by most operators to the trocar, and it may be that a better knowledge of the operation gives it its apparent advantage. Dr. Roberts professes a decided preference for it. He would have the receiver exhausted of air beforehand, the penetrating instrument introduced slowly, and the whole so arranged that the operator can see the first drops of fluid that escape from the sac, and thus avoid penetrating too deeply. Dr. Roberts considers the "Point of Puncture" at some length, and holds that one of two should be preferred— "the fossa between the ensiform and the costal cartilages of the left side, and the fifth intercostal space near the junction of the sixth rib with its cartilage," and after some reasoning he adds: "Therefore I should tap in the former position as a rule," i.e., the fifth left space, "reserving the latter" left xiphoid fossa "for special cases, where there was some indication for making an exception."

In such a case as that in the New York Hospital the puncture might be made almost anywhere in the front of the chest below the fourth rib, for the sac extended as much to the right as to the left, while the heart kept its proper position. In such a case, "Dr. Robert's suggestion to tap on the right side of the sternum in the fifth space, about four and a half to five centimetres from the edge of the sternum," would be safe and proper. But *such* cases are extremely rare, if indeed another gallon sac has been seen by anybody. But any pericardial effusion that produces distress enough to raise the question of tapping will be very exceptional if it does not extend to the right as well as to the left, and how far it is easy to ascertain.

You are too fresh from your anatomy to make it needful that I point out the danger of wounding the internal mammary artery.

Tapping the Pericardium—Recovery.—It was in the Leeds Infirmary, service of Dr. Allbut, case reported by the house physician, and published in the *Lancet*, January 27, 1882. The patient, twenty-two years of age. He had had rheumatic fever six or seven years before. He was admitted December 23, 1883, with pains in his joints, not undoubtedly rheumatic. Salicylate of soda did not relieve him; seemed to produce some delirium (20 grs. every two hours) and was discontinued. Had pleurisy on the 23d of January; he had been steadily improving for the last four or five days. This morning he is pale and breathing rapid; not in any great pain, but a little discomfort in the breathing; cardiac dulness much increased. It begins at the third rib and extends downward to the sixth, and *is much increased to the right of the sternum*, distinct friction fremitus felt over the upper part of dulness, friction sound heard over upper half of dulness, the systolic portion of which was double; pulse 130, evening temp. 100.6°.

Feb. 2.—Pericardial effusion rather increased. The præcordial region has a full distended look, no intercostal sinking. The absolute dulness begins now almost at the second rib, and extends down to the seventh, laterally on the fourth rib it measures eight inches; heart sounds obscure and distant; no friction sound now. To-day for the first time the epigastrium is full and tender, and dull on gentle percussion. The edge of the liver on the level of the umbilicus. Temperature, 100°; pulse, 128.

Feb. 7.—It is evident that he cannot hold out much longer if present condition continues. A fine aspirator needle was introduced in the fourth left space, two inches and a half from the median line, directed upward and backward until a cavity was clearly entered at the depth of an inch and a half. Only about an ounce of bloody turbid serum could be drawn off. It was not till the 11th that the patient began to mend. He continued to improve, and left the hospital May 18.

It may be that the withdrawing of an ounce of fluid in this case relieved pressure, and so favored absorption. Every physician of large practice has seen this kind of relief in pleurisy. A patient has one pleural cavity filled, distended with serum. He tries blisters and diuretics in vain. I have tried all approved means for two months without making any favorable impression. A portion of the fluid is then drawn off, and it need not be a large portion; directly the same agents that were in operation before become effective, and the fluid is absorbed in a few days. That Dr. Albutt's patient was relieved by absorption needs no argument. Whether the absorption would have occurred without the tapping may be a question. The statement I have quoted, "It is evident he cannot hold out much longer if the present condition continues," has an important bearing on this point. If there are cases in which the removal of a part of the fluid will determine the absorption of the remainder, such cases will be a sort of counterpoise to those that require repeated tappings.

Signs of Pericardial Adhesions.—On this topic there has been a great deal of fine reasoning, and some good observation, but observation and apparently good observation varies widely in results. Indeed the cases must vary one with another. I have already shown specimens in which the pericardium is closely and firmly adherent to the heart, so that the pericardium must make every movement that the heart makes, and have shown how this kind of adhesion produces a twisting movement in the diaphragm. I have also shown you specimens in which the adhesion was not an adhesion, though it was at one time. The plastic matter had been drawn out into threads half an inch long, by the interposition of serum. The serum remaining more than the usual time the plastic matter may lose its adhesiveness, and the connection be kept up by threads only. Again you have seen when close adhesions are breaking down, the last stage is marked by a multitude of very fine threads running from the pericardium to the heart, and both cleared of false membrane except at the points of attachment of these threads. In either of these cases the heart has its natural play without dragging the pericardium after it. In the case of dissolving adhesions you will rarely understand that a set of symptoms indicating adhesion may have lasted one or more years and then have gradually disappeared.

Still again you have seen instances of close adhesion, perhaps equally close, while the hearts varied greatly in the quantity and quality of their muscular fibres. Here, for example, is a *cor bovinis*, hypersarcosis cordis; and here is one that is not hypertrophied, but of natural size; the pericardium is closely adherent to each. The power of one is double, perhaps triple, that of the other. This is not, perhaps, felt in the vessels because there is a grave obstruction at the aortic valve but it is exerted in the chest and felt in its walls and will give signs of its presence that the heart of normal size cannot give. Here is still another; it is, perhaps, under size, but I will not make a point of that; it is clearly not enlarged, but it is yellow in color, in consistency flabby and soft. It is an instance of the oily degeneration of the muscular fibres of the heart. The pericardium, you see, is closely attached as in the others. But you would expect during life no signs of this adhesion whatever.

There is another point in this history which is of great importance. It is said that when the serous pericardium is inflamed the diseased action shows a disposition to penetrate the whole thickness of the sac, and appear on its outside, and that the result of this action is often a firm adhesion of the pericardium to the inner surface of the sternum and of the costal cartilages. That this does occur cannot be denied, but as to its frequency I may state that most of numerous specimens by means of which I have demonstrated to you the various phases of pericarditis were removed from the body by my own hand or under my immediate direction, and that when the sternum has been separated from the clavicles and the cartilages from the bony ribs the raising of these was an easy work; a few broad, careless sweeps of the knife have sundered the scanty connections of the sternum and its appendages to the parts within. It has rarely been necessary to make careful dissection to separate the pericardium from these parts. But when such adhesions exist, they certainly have very telling effects among the signs of pericardial adhesion.

It comes then to this: the most marked of these signs will be observed when the adhesions are close, when the heart is hypertrophied and strong, and when strong attachment exists between the anterior of the pericardium and the inner face of the sternum and cartilages, and that there will be feeble external signs when such expericardial attachments do not exist and the heart is of usual strength, and none at all when the heart is feeble, whether the expericardial attachments exist or not.

In consideration of these points the following can be understood:

In a case in which it was all but certain that these adhesions existed, and in which they were found to be "close and universal," after death, Dr. Wm. Pepper sought for such evidences as have been supposed to prove it. "But in fact they were entirely absent. The impulse of the heart was diffused and feeble, and unattended with thrill. There was no recession of the intercostal tissue during the ventricular systole, nor any diastolic collapse of the jugular veins, nor recession of the epigastrium. And this was the case despite the fact that there existed those external adhesions between the pericardium and the chest wall in front, which seem in some cases to render the above mentioned signs of pericardial adhesion more evident. This case must be regarded as another illustration of the fact that while, when these signs are present, the existence of adherent pericardium may be assumed with great probability. They may all be absent in cases of complete adhesion."

Another observer, writing in the *Union Medicale*, relates a case apparently of tapping the pericardium and states that "the doubtful position and value of this operation makes the case of interest. The patient suffered from a pericarditis with effusion, produced by exposure to cold. The symptoms of pressure were severe; a puncture was made and a little fluid removed. A month later the patient had double pleurisy. He finally recovered with pericardial adhesions." (The *Med. Record*, Jan. 6, 1883.)

Dr. Sibson, in "Reynolds' System of Med.," who has written elaborately if not lucidly on these signs, says: "The discovery of adherent pericardium during life is in some cases impossible; in some doubtful or difficult; but in others, and these are among the most important cases, its existence can be ascertained during life, on reasonable and well-ascertained grounds.

"When the adhesions are partial, or when the heart, though completely adherent, is small, is not bound by external adhesions to the anterior walls of the chest, and is covered to the natural extent by the lungs, their expansion being free and unconstrained, then the varying relation of the heart and lungs to the chest is quite natural, and the diagnosis of the adhesions is impossible. If the adherent heart be enlarged, and is not attached to the lower half of the sternum and the cardiac cartilages by combined pericardial and pleural adhesions, so that the active or automatic and passive or respiratory movements of the heart are scarcely or but little, interfered with, the inspiratory expansion of the lungs is freely permitted, and the diagnosis of the adherent pericardium may be difficult, obscure, or even impossible.

"When, however, the heart is, as usual, enlarged, being often affected with valvular disease, the adhesions may be short, fibrous, and binding; and the front of the organ may be fixed to the two lower thirds of the sternum and the adjoining cartilages by pleuro-pericardial adhesions, so that the automatic and respiratory movements of the heart and the inspiratory expansion of the lungs are restrained: thus the discovery of the adhesions, during life, may generally, in such cases, be made by a careful study of the physical signs; its diagnosis being the more certain and easy in proportion as the heart is more enlarged and more firmly fixed to the anterior walls of the chest."

Puncture and Suture of the Heart.—Dr. Roberts read a paper at a meeting of the Philadelphia College of Physicians, in which he cites some instances in which the heart has been punctured when the purpose was to tap the pericardium, and claims that these accidental punctures did no harm, and calls attention to Dr. Westbrook's paper (published in the *Medical Record*, Dec. 23, 1882), in which he describes an intentional tapping of the right auricle to relieve a heavy congestion, and urges, that if a few drachms of blood taken from the heart will give as much relief as the same number of ounces taken from the arm, it is a great economy of blood.

He thinks also that important results will follow Dr. Block's experiments on animals, which show that in them not only can the pericardium be opened to remove clots of blood, but that opening the right or left ventricle and entire compression of the heart for the application of suture can be supported by rabbits for several minutes. Even if the cardiac pulsation and breathing stop during the operation, death, he asserts, does not necessarily follow. (For Dr. Block's experiments see *Jour. of Med. Science*, Jan., 1883, p. 274; *N. Y. Medical Journal*, March 17, 1883.)

Pericardial Puncture and Incision and Drainage.—At a meeting of the Royal Medical and Chirurgical Society, Dr. Samuel West reported the case of a boy, 16 years of age, who had a large pericardial effusion. The symptoms became so urgent that paracentesis was performed. Pus was obtained. Three days later paracentesis was again performed, and subsequently the pericardium· was laid freely open, evacuated, washed out, and a drainage-tube inserted. The boy recovered completely in five weeks, having had in the meantime an attack of urticaria. The place selected for the operation—puncture—was the fourth intercostal space, immediately behind the left nipple. The amount of fluid obtained at the first tapping was fourteen ounces; by the incision, two quarts. There was a peculiar epigastric prominence noticed before the paracentesis, which disappeared after the operation. The pulsus paradoxis was constant till the free incision was made, and ceased immediately after that.

Dr. West then gave a short account of the only other recorded case of incision of the pericardium for purulent pericarditis, by Prof. Rosenstein, of **Leyden**, which also recovered.

Dr. West then gave a history of the operation. It was first suggested in 1649. It was first practised in

Barcelona, in two cases. In 1841 there was a remarkable series of cases in an outbreak of scurvy in Russia, in which the pericardial effusion was mostly blood. Nine were operated on, and six recovered. In 1854 Trousseau's essay was published upon some cases of his own and of Mr. Aran, which revived an interest in the subject. In 1866 Dr. Clifford Allbutt introduced the operation into England, and it was performed by Wheelhouse and Mr. Teale. Rosenstein, in 1871, made the advance, in opening the pericardium by incision with drainage. A list of recorded cases, including some hitherto unpublished, was given in a tabular form, making 79 cases in all. . . .

Phthisis and pleurisy were associated with 23; rheumatism with 11; scurvy with 9; general dropsy with 5; injury with 3; in 12 there was no associated disease. The amount of fluid evacuated was in 46 cases less and in 38 more than a pint. The larger quantity was in the scorbutic cases, and from one of these about ten pints was obtained. Dieulafoy selected the fifth left space, about an inch from the sternum, as the safest point for puncture. Only one case is recorded in which the operation was fatal. (*Am. Jour. of Med. Sci.*, July, 1883.)

Penetrating the Right Ventricle.—On the presentation of Dr. West's case of pericardial puncture and incision, Dr. Hulke said that he considered it advisable to dissect down carefully to the pericardium before any incision was made, and if a trocar and canula were employed, he advised a very careful use of them, and that the trocar be frequently withdrawn to form an opinion of the parts reached. He had himself, after medical consultation, in a case that was believed to be one of pericardial effusion, once inserted a trocar and canula somewhat boldly, and the withdrawal of the trocar had been followed by a jet of blood which gave him great alarm, but happily relieved the patient. A subsequent *post-mortem* examination showed him that he had punctured the right ventricle, and that the case was one of universally adherent pericardium. (*Am. Jour. of Med. Sci.*, July, 1883, p. 264.)

Dr. Partzersky is reported (*Am. Jour. Med. Sci.*, April, 1883) to have brought before the Moscow Physico-Medical Society a very interesting case of pericardial effusion, treated by repeatedly performed tapping, and finally by incision of the pericardial sac, with subsequent drainage. These are the author's views: 1. In a vast majority the operation—that is, puncture and aspiration, and, if they fail, subsequent incision with drainage—is not attended with any danger. 2. It brings rapid relief, and its palliative usefulness is admitted. 3. In the absence of such complications as tubercles, cancer, organic changes of the heart, etc., the operative treatment of non-purulent pericardial effusions may prove successful in the majority of cases. 4. In purulent pericarditis an early operation is justifiable, in order to prevent dilatation and fatty degeneration of the heart, which generally supervenes very rapidly.

Cancer of the Œsophagus Producing Pericarditis and Pneumo-Pericardium.—There had been a steadily increasing difficulty in swallowing, in a woman of 43, for many months. There was much regurgitation of food. At length the to-and-fro sound of pericarditis was recognized. Six days after a *churning splash*, a resonant percussion indicated that air had entered the pericardium. She died two days later, and it was found that a cancer —(epithelioma?)—had made extensive ravages, and had penetrated the posterior of the pericardium so as to make an opening that admitted the point of the finger.

Dr. Begbie reviews the history of pneumo-pericardium, and recognizes three different ways in which it may be produced: 1. It may be secreted. 2. It may result from decomposition of fluid in the pericardium(?). 3. It may by perforation be admitted, whether that is caused by disease or injury.

That a gaseous matter may be secreted under such conditions I believe is possible, both from the reliableness of those who have recorded such cases and from one that I have myself seen. Dr. Stokes has seen one such case, and Laennec thought it was a common occurrence in the dying, and he has given accurately the physical indication of it. If this mode of production is admitted, then the gaseous decomposition of the contents of the pericardium is rendered doubtful. Bricheleau did indeed meet with one case of pneumatosis of the pericardium, and on postmortem examination the pus found in the cavity was found to be very fetid. But even in such a case it is as easy to suppose that the gas was carbonic acid gas eliminated from the blood-vessels as it is in the intestinal tube, and taking the odor of the fetid contents of the sac in the same way that an abscess in contact with an intestine acquires a fæcal odor without fæcal admixture, as it is to admit gaseous decomposition. Into the third class falls a case mentioned by Dr. Graves, in which a communication between the stomach and pericardium was produced by an abscess of the liver which opened into both, and one by Dr. Flint, in which the patient was stabbed with a knife, which made a slight opening in the pericardium; one by Dr. Walshe, in which a juggler's knife, in the attempt to swallow it, was arrested in the œsophagus and wore a way into the pericardium; and that reported by McDowel, in which a cavity of the left lung had penetrated the pericardium.

The Signs of Pneumo-pericardium.—The few physicians who have reported cases of pneumatosis of the pericardium are entirely agreed regarding the diagnostic signs. It appears that no case has been yet recognized of air or gas in this cavity in which there was not fluid of some kind in it at the same time, and hence the most characteristic feature of the case—the churning, or splashing, or water-wheel sound. This is heard in systole, and with every contraction. In some this sound has been heard at some distance from the body, in others only by actual application of the ear to the præcordial space; but in all it was distinct and unmistakable. It has been. called tympanitic. Stokes denominates it *bruit de pot fêlé*, Laennec, *bruit de fluctuation*.

In all there has been resonance on percussion over the heart, which changed place with changed position of body.

Dr. G. Tilling reports a case of traumatic hæmato-pericardium. The patient was caught between two wagons; was at first insensible, but recovered consciousness and walked home. He was spitting blood when first seen, twenty-four hours after the accident: The apex-beat of the heart was perceived by the eye, difficult to feel, and diffused. The area of dulness extended a finger's width beyond the right edge of the sternum. The heart sounds were marked, by various indescribable murmurs in this region, but were clear and distinct at the second intercostal space, no air in either pleura or pericardium, no fracture, pulse slow, 54 to 76. Was the diagnosis of hæmo-pericardium, as against hydro-pericardium sustained by the facts as reported? The patient recovered. (The *Medical Record*, Sept. 23, 1882.)

Dr. Tilling thinks that the following symptoms justify him in believing that there was an effusion of blood into the pericardium in this case of injury of the chest which did not break any ribs. Cardiac dulness extended over the right border of the sternum. The apex-beat could only just be felt, but could not be located. The heart sounds were marked by a variety of sounds, blowing, rubbing, and splashing, except in the second intercostal space, where they were clear. The peculiar splashing sound would seem to indicate a "partial" pneumo-pericardium. A similar sound, he says, was recorded by Morel Lavallée in a case of actual hæmato-pericardium, as proven on inspection. But Tilling's patient recovered. On the third day the rubbing and splashing sounds disappeared. In a case reported by Billroth he says, *pericarditis* occurred from a blow, but the *presence of hæmato-pericardium is not mentioned.*

LECTURE III.

ENDOCARDITIS.

IN my opening remarks on pericarditis I gave you the boundaries of dulness produced by a healthy heart, that you might know when, by pericardial effusion or other cause, those limits were exceeded. So now, as we begin the study of endocarditis it may be interesting, and perhaps more interesting than useful, to state to you the exact position of the four valves of the heart, their relative position, and their relation to the walls of the chest. This information will be clinically important so far as it relates to the aortic and pulmonary valves ; but in the case of both auriculo-ventricular valves it is of less practical value, because the sounds produced in them are not heard directly over them, but over the part of the heart to which the vibrations are conducted by the apertures of the valve. Yet even in regard to them it cannot be regarded as useless knowledge.

After the paper on "Auscultatory Percussion," already referred to, was prepared and had been sent to the printer, I began (June 27, 1840), in association with the late Dr. Swett, the studies to be here reported. A careful record was made at the time of what we did, which has never been published. Indeed, I have never seen a report of any similar experiments by anybody. We repeated the trials on different bodies till we were assured that the results were fairly reliable, choosing those in which there was during life no evidence of cardiac disease.

We used the same kind of penetrating instruments that were used in defining the limits of the heart, i.e., sharpened steel knitting-needles. The first step was to raise the trachea and cut it just below the cricoid cartilage, introduce a compressed cork, and tie the cork firmly in. The object of this was to preserve the relations of the heart and lungs by preventing the escape of air from the latter, when the chest should be opened ; and also to prevent any possible change in the position of the heart from the same cause. Then we laid bare the sternum, cartilages of the ribs, and part of the ribs in the præcordial region.

• A needle was forced through the sternum one line to the right of the median line of the body, and half an inch below the line that would unite the points when the lower edge of the cartilages of the third ribs join the sternum. The needle was introduced at right angles to the plane of the body. It was found to have passed just outside the aorta to the right on the level of the free border of its valve. This trial, then, states that the whole aortic valve is below the inferior edge of the third rib at the sternum, half an inch, and that it does not extend more than half a line, or the twenty fourth of an inch to the right of the middle of the sternum.

A needle introduced at the same point in another body passed into the top of the right ventricle, barely escaping the ventricular septum, and on to the right of the aorta, and, entering the right auricle, touched the left attachment of the tricuspid valve. A little deeper it passed less than a line to the right of the right limit of the mitral opening. Here, then, it is shown that the aorta at its origin is on the left of the point chosen, i.e., half an inch below the third rib and one line to the right of the middle of the sternum ; and that this point indicates almost exactly, if we measure from left to right, the end of the mitral opening, and the beginning of the triscuspid, of course on different planes.

A needle introduced in the third left intercostal space,

equidistant from the third and fourth ribs, and in contact with the sternum, penetrated the posterior aortic cusp midway between its free border and its attachment, also at a central point from right to left. The same needle penetrated the anterior curtain of the mitral valve, passed through this opening half an inch from its right extremity. In one trial the needle introduced at the same point left the middle of the base or attachment of the aortic valve five lines above it, but it pierced the curtain of the mitral valve and passed through the opening at the central point, leaving eight and a half lines on either side of it. This heart was that of a man who died of typhoid fever, and the heart was in the wet-rag condition of Louis, everything "flattened out." Thus it appears that the aortic and mitral valves are in the same vertical plane; commonly in the same horizontal plane, but not always —the aortic directly in front of the mitral, but sometimes above it.

A needle penetrating the sternum at a point one line to the right of its middle and one inch five lines below the junction of the third rib and sternum, lower edge, passed into the right ventricle, and penetrated the tricuspid curtain just at its attachment, and midway between its extremities, entering the muscular tissue that forms the lower, or better, the right lip of this opening.

The pulmonary valve we found situated further to the left than the aortic, and only a line above it. Measuring from the needle that was inserted in the middle of the intercostal against the left edge of the sternum, we oftenest found the middle of the base of the aortic valve five lines vertically inward; while the middle of the base of the pulmonary valve was six lines to the left and then six lines vertically upward measuring still from the same needle.

From these trials it appears that the aortic valve is

half of it under the sternum and half of it under the lower part of the left third rib—that the pulmonary valve is almost wholly under this rib, leaving only about a line of its width under the sternum—that the mitral opening begins half an inch below the lower edge of the third rib at the middle of the sternum, and extends to the left and a little upward to the cartilage of the third, in a state of collapse eight and a half lines from the left edge of the sternum—and that the tricuspid begins at a point less than a line to the right of the right limit of the mitral, and extends almost directly downward, being wholly under the right half of the sternum.

The measurement of the mitral opening, as just given, corresponds to a circumference of thirty-four lines, and this to a diameter of eleven and one third lines when rounded and in action. This would diminish the lateral space occupied by this opening from seventeen lines to about an inch. This measurement was recorded only for the purpose of saying that the needle was at the exact middle of the opening. The measure of the other orifices was not taken, as the object of these trials was to ascertain position, not capacity.

Knowledge is always worth having, but as I have already said it is more practically important to know the location of the aortic and pulmonary valves than that of the auriculo-ventricular. In case of the first we listen directly over and not far from them for the evidence of unhealthy changes in them, but the indications of such changes in the mitral valve are often not obtainable when the ear is applied exactly in front of it, for the aortic and pulmonary artery are between the surface and the valve, and more than half the time there is blood in the ventricle coming between the valve and the ear in addition. But this valve is connected with the walls of the heart by its own tissues, and by the tendinous cords and fleshy columns, which are good conductors of sound. We therefore get the sounds produced here most clearly over a point where an anterior fleshy column is incorporated with the cardiac walls. This is not at the apex exactly, but about three quarters of an inch nearer the base, but for convenience of expression and shortness we almost always call it "listening at the apex."

In the tricuspid valve the anatomical arrangements are very similar, but we can listen *over* it with better chances of perceiving morbid sounds, because the right ventricle is superficial, but we also listen just off the sternum in the fourth right intercostal space.

Dr. N. Werp* reports an instance of constant diastolic murmur for four weeks before death. On inspection there was found a patch of recent endocarditis on the auricular surface of the aortic half of the bicuspid (mitral) valve. There were also traces of old endocarditis at the bicuspid with consecutive valvular aneurism, endarteritis deformans of the aorta with dilatation of its ascending portion, excessive hypertrophy of the left ventricle, commencing pericarditis. The little aneurismal pouches on the mitral valve were turned with their cups facing the auricle, and were also slightly roughened. There was no stenosis of the venous opening into the auricle. Here, then, there was no valvular lesion (?) in the ordinary clinical sense of the term, yet there was a persistent diastolic murmur at the mitral opening.

Now we are ready to pursue our study of endocarditis. I have gone into this explanation of the action of the valves, and the effects of their not working properly because endocarditis is a deformer of valves.

When inflammation occurs in the endocardium, we are not sure that it affects the whole extent of the endocar-

---

* The *Med. Record*, Sept. 23, 1883.

dial membrane. We are sure that it occurs about the valves. The tendency of persons who write about diseases of the heart is to confine the inflammatory action to the valves. Now, it is not necessary that the inflammatory action should be confined to the valves even from what we see. If there is an exudation upon the border of the ventricle, for example, on that part of the endocardium, it will be washed off by the current of blood that is continually flowing, and flowing with a good deal of force, through the heart. You ask why it would not be washed off from the valves? Because it holds fast there. The effusion is between the folds that make the curtains of this valve on the attached surface of these folds, and it cannot readily get out. Then there is a form of inflammation called ulcerative endocarditis that attacks any portion of the endocardium. I will explain that further on. The point, then, I make is, that because the results of the inflammation are found only in and about the valves, it does not follow that the whole endocardium may not have been the seat of disease.

The question occurs, what changes take place in the endocardium during the inflammatory process? I do not know whether it becomes red, except by analogy. Nobody has ever seen the endocardium at the commencement of endocarditis. We cannot say, then, that we know the membrane is red. We can infer that it is probable, because membranes that are inflamed anywhere else, so far as we know, are red at the beginning of the inflammatory process, becoming less and less red as the inflammation exhausts itself. Roughnesses are not commonly made (except in particular kinds of inflammation) upon the body of the endocardium, but here at the valves. You may expect from endocarditis, first, a thickening of the valves; and second, a growth, either upon the valves or near them (but this is not constant) of little eminences, little granulations, you might call them, which are described as being firm at the base, and by microscopical examination found to be composed of connective tissue at the base, but of new and soft cells at the top, so that the base of these little granulations is firmer than the apex. One of the most important facts, perhaps, in the whole history of endocarditis is the formation of little granulations upon these valves, causing the blood to precipitate its fibrine upon these rough surfaces. You are, perhaps, aware that if, in an animal, a thread be run through a vein or artery so as to pierce it about its middle, the fibrine will collect upon that thread and form a clot, a considerable little ball; or it will be swept in the direction of the circulation. Any rough body in an artery, or in the heart, or in a vein, will cause a coagulation of blood about it, the fibrinous portion of blood making the clot. Well, here this roughened surface has become encrusted, so to speak, or covered by a coagulation of fibrine, having some blood-coloring matter in it to give it a red color. Now, this is the worst thing that can happen in endocarditis, for the reason that these concretions, these aggregations upon any rough surface of the heart, are apt to be washed away and carried into the organs of the body—into the brain, not infrequently; rarely into the liver, because the liver circulation from the heart direct is feeble, being supplied chiefly by the portal circulation. But the brain lesions are the worst, and this embolism carried from the heart valves most frequently, or from any portion of the heart, to the arteries of the brain, will frequently produce apoplexy—of course, a pretty grave matter. The principal points to be noted with reference to the effect of endocarditis are: first, thickening of the valves by an effusion between their folds; and, second, these granulations growing up in different parts of the endocardium, mostly upon the valves which have lain traps and caused a deposit of the fibrine of the blood upon them, making what we usually call vegetations, and these are constantly apt to be washed off and carried into important parts of the body and do serious harm.

Now, the material that produces thickening of these valves is not exactly like ordinary false membrane. The inspection of it by the microscope, when it is recent, shows that it is a transparent, opaquish fluid, filled with new cells, not of the kind that we refer to pus, but those that are capable of combining and being propagated into fibrine. The false membrane met with in endocarditis is made in a different way from ordinary false membrane, with which you have already become somewhat familiar. It has no property of its own. It is coagulated into fibres by the addition of acetic acid; at least, Virchow says so, and he compares it with mucine.

This product may, at first, swell the valve considerably, and afterward contract by the absorption of its fluid part, and then it may become organized with the two folds of the valves. The result of this will be a thickened and clumsy valve. And as this new material contracts it contracts the valve, the valve is shortened so that it cannot reach its fellow in the middle of the artery, and it is very much stiffened and hardened; it is of the consistency of leather sometimes. The defect described is the cardinal one. The valves are thickened, they are shortened, they are stiffened, they cannot perform their office. The result of this is obstruction, and again, regurgitation, one or both, as may occur in a particular case.

Now, then, with reference to the recognition of these changes as they are coming on. You sometimes can recognize them and sometimes cannot. Endocarditis is very apt to occur with pericarditis, and it is often difficult to determine whether there is endocarditis or not.

There may be a grade of inflammation of the endocardium and of the valves that will not produce this mucine matter, but merely œdema. The valves may become thickened by becoming bags of water. Now this is a form of disease that is sometimes recognized, but not by the rational signs. To recognize an endocarditis by rational signs is impossible, and perhaps this is a good reason why it was not recognized during life till auscultation became an art, and endocarditis may, in that sense, be called the child of auscultation. When this œdematous effusion occurs, the valves are obstacles to the free flow of the blood, and produce a murmur. If, then, you have listened at a particular time to the heart and found no murmur of any kind, and you have symptoms of a very vague kind, and you suspect endocarditis, if you listen again you may possibly hear an endocardial murmur, and it is very likely to be at the aortic opening.

Relating to endocarditis, the matter of vegetation, commonly so called in this country, is of interest. I told you that endocarditis is sometimes attended by the formation of little granulations growing out of the endocardium, firm at the base and soft at the top, not so large as a pin's head; and that these, though of very diminutive size, are capable of being the nucleus around which the fibrine of the blood will coagulate, or rather will be gathered, and become of visible form. What we call vegetation, the German writers regard as an accretion, but we are both correct in regard to its nature, viz., that it is substantially the fibrine separated from the blood in the same way as it is separated when a thread is run through an artery or vein. Any cause that produces roughness, as an ulcer or any irregularity, is liable to be surrounded by this accretion of fibrine. While we rarely see a membranous exudation in endocarditis upon the body of the heart, on its inner surface, that can be easily accounted for, under the supposition that the exudation has thickened and fallen off by the continual washing of the surface by the current of blood. In this specimen you see some deposit of fibrine on the three aortic curtains. And here is that rather interesting—I don't know but I ought to say ornamental— formation of granules, forming a semicircle on the free portions of the valves. This, however, illustrates atheroma.

One of the dangers, as I have said, of endocarditis is the washing away of these deposits into the general circulation, and, finally, the lodgment of a mass of this kind in an artery, through which it cannot pass. Endocarditis of itself is almost never fatal. It accompanies pericarditis in a great majority of cases of pericarditis, and for the most part, when so associated, depends upon the same cause. Yet endocarditis can occur even in the acute form, and frequently it does occur in the chronic form, without any rheumatism preceding it, without any of the diseases that are apt to be associated with pericarditis, as Bright's disease, scarlet fever, purulent inflammation of the body, septicæmia. I have sometimes, by exclusion, been able to say that a certain person has endocarditis. I form such an opinion when I find a certain amount of febrile action, a little elevation of temperature, a little increased frequency of the breathing, and a short, irritable, quickened beat of the heart. That short, irritable beat will produce in the ear the same sound that I referred to in describing pericarditis; when it was first noticed it was regarded as diagnostic, but it has no such value. It is just that metallic ring which you get by applying the palm of the hand to the ear to exclude the air, and then striking on the back of it. It is the irritable beat of the heart against the inner wall of the chest that produces it. You should be able to exclude everything that would be likely to produce a febrile movement, and perhaps you may, in half the cases, find that the diagnosis of endocarditis is verified by after indications. During the progress of endocarditis there are frequently no indications of its presence except pericarditis. You assume, usually, at once, when you diagnose acute pericarditis, that there is endocarditis, but of itself it does not give distinctive signs except in a few cases. When, for example, you have, say rheumatism, or Bright's disease, and have listened to the heart on a certain day and have found nothing unnatural, you find the illness is increased a little, the heart beating more frequently, the respiration short and frequent, you listen again and you get an endocardial murmur. It is probably soft. It may be at the aortic opening; it may be at the mitral. Supposing it to be in the interval of the heart's action, during the period of repose, it is produced by one of two things: an œdematous valve, or the little granulations I have been describing forming upon the surface that the blood has to pass over. In a few instances you may get that aid to diagnosis, but in the majority you do not. Then, again, there may be an œdematous swelling of the aortic valve; the two folds that constitute the valve are swelled out and partly fill the passage through which the blood should flow. A soft murmur is sometimes produced in that case, but as it is only sometimes it cannot be relied on as a diagnostic mark, except when you have listened a few days before and found no murmur, and now, with indications that there may be endocarditis, you do find one, then it is sufficient; but, in practice, neither the rational nor the physical signs of endocarditis are sufficiently marked to enable anybody to make a certain diagnosis without one of these facts that I have just now referred to. In general, endocarditis comes during its acute period without producing any symptom, only that short irritable beat of the heart and the increase in the frequency of the respiration. The rational diagnosis is entirely vague. The physical diagnosis also is very uncertain. But the after diagnosis is less difficult—you may lock the door after the horse is stolen—the after diagnosis is certain. The deposit in these valves, if it is anything more than œdema, is pretty sure to produce changes in the structure of the valves. In the first place, they must be thickened, for there is new matter deposited in their structure; in the second place, this new matter undergoes organization. It behaves like a false membrane, and shortens, and the valve is shortened, so that when it attempts to fill the opening that it is appointed to fill it cannot. There is a hole left in the middle. The valve may sometimes grow hard and stiff, like leather, and then can do very little in preventing the return of blood, but a good deal in preventing the outflow of blood. When these changes occur, even in a moderate degree, you will get a murmur. It will very likely be of a harsh kind. It is not however certain to be that at first. It will grow to be a harsh murmur, in all probability, in a few months. The period of time, however, at which you can recognize the murmurs will be from three weeks to three months after the disease has passed, after the time when your diagnosis would be of any practical value. But the thing stands so. You have to take it as it is. You can pretty safely say, after two or three months, this patient has had endocarditis. But as it is so frequently associated with pericarditis, and as pericarditis is easy of recognition, and as, further, the treatment of the one condition is the same as the treatment of the other when they are associated, this lack of power of diagnosis is not disastrous. It is only important when the disease occurs alone; and that it does occur alone we have pretty good evidence in the great number of persons that we see with valvular disease who have never had rheumatism, never had Bright's disease, never had any of the affections that are so apt to precede pericarditis and endocarditis. They have enjoyed, as they believe, good health, and cannot point to any time when they have had symptoms that would lead us to think that they had had heart disease; yet the valves have become thickened and are defective, just as they are in the acute form of the disease. There is then undoubtedly a chronic endocarditis, that only shows itself by the physical changes that are produced, or rather by the changes that are produced in the sounds of the heart. Dr. A. J. Harrison,* after detailing three cases of endocarditis, occurring without disease of the joints, urges that the occurrence of the disease in this way as a primary affection is not very rare, especially in those predisposed to rheumatism. He is inclined to regard endocarditis as one of the regular symptoms, rather than one of the frequent sequelæ of rheumatic fever. Haskins believes that the disease is actually the initial fact in every attack of rheumatism, and that its presence can always be substantiated by physical signs. Others, while acknowledging the constancy of roughened heart sounds in rheumatism, do not consider that this sign possesses the significance Dr. Haskins assigns to it.

Dr. Harrison also read before the British Medical Association * the history of four cases of primary endocarditis. The president, Dr. Allbutt, said that cardiac disease should not be taken as a test of severity of rheumatic fever. Endocarditis was frequently the first event in that fever. It is not uncommonly the only event. He referred to cases, and especially to one in which endocarditis occurred without other local manifestations.

Dr. Balfour said that every case of systolic murmur

---

* N. Y. *Med. Jour.*, March, 3, 1883.
* The *Medical Record*, Aug. 26, 1883.

should not be considered a case of endocarditis, because a certain amount of rheumatism might occur without any endocarditis; then he had not seen a case of rheumatic fever in which there was not a systolic murmur. In fever of all kinds, especially in feeble persons, a systolic murmur could almost invariably be detected. Besides he thought that every case of friction should not be regarded as a case of pericarditis.

Dr. Ashley remarked that he had frequently observed endocardial and pericardial murmurs in connection with slight attacks of tonsillitis. A fact long ago observed was that tonsillitis had a casual connection with rheumatic fever, the same as influenza.

The relation of these changes to the structure of the heart itself will be for consideration hereafter. You observe, then, you have not a very certain diagnosis. Only when you have listened and heard no endocardial murmur, and then, a few days after, with symptoms of illness, do find one, are you enabled to make a diagnosis. And then you have to be certain that it is an endocardial,—not an exocardial murmur. The statement I have made with reference to the breathing will aid you in making the diagnosis.

Dr. Putnam-Jacobi* reported to the N. Y. Pathological Society the case of a child thirteen years of age who gave proofs of a prior attack of both left and right endocarditis, which probably occurred during an attack of rheumatism a year before her death, and also deposits by an endocarditis of the right side that were recent fibrine particles from the last deposits which had been swept into the pulmonary current and produced embolism, and in turn she thinks produced pneumonia, and this again pleurisy,

---

* The *Medical Record*, March 17, 1883.

both of which existed. Colomeatte * describes five cases of acute endocarditis of the tricuspid valve, and says that this inflammation may fall on the tricuspid or sigmoid valve alone, or on both at the same time; that it may be perforating at one orifice while it causes vegetations at the other. It is found in infants, in youth, in old age, and in both sexes. From a sixth case he infers that endocarditis may affect only the right wall, or the vegetations may be limited to the right auricular appendage.

These vegetations were composed of embryonic connective tissue, the elements of which were for the most part in a state of fatty degeneration. They were easily torn and frequently caused pulmonary emboli.

Dr. Ord, St. Thomas's Hospital Reports,† relates a case in which shivering fits occurred daily for a fortnight, then every other day, and then every third day. After admission the patient had daily rigors for five weeks, when he died. There was on admission a marked presystolic thrill over the impulse, a systolic murmur at the apex conducted into the left axilla, and a fainter and apparently independent systolic murmur over the aortic valve. The arteries were everywhere much thickened. The liver and spleen were large, and the spleen was tender for one week.

At the post-mortem examination there was found pericardial effusion 15 oz., but no lymph; dilatation of both ventricles and hypertrophy of the left. The posterior set of chordæ tendiniæ of the mitral valve were ruptured, their free ends clubbed and covered with black clot. The endocardium showed a white tract, where the free end would have played against it. Bright's kidney, large white; large liver.

---

* The *Medical Record*, Feb. 24, 1883, from London *Medical Record*, Jan. 15, 1883.
† *Am. Jour. of Med. Science*, Oct. 1882.

Prof. Leyden* calls attention to the frequent resemblance of the temperature curve to that of intermittent fevers. This fact has been many times observed by others. He makes four groups of cases clinically. The first, those cases in which the endocarditis forms part of a pyæmic or septic process. This is best known in connection with puerperal septicæmia, in which ulcerative (infectious parasitic) endocarditis is not uncommon. It is also met with sometimes in septicæmia and phlebitis following injuries. Leyden observed one such case connected with abscess of the liver. As in these cases the affection is general, the rigors may be a symptom of general sepsis, quite as much as of the endocarditis. In the second group he places those cases which are marked by a more or less intense and irregular pyrexia and erratic rigors. Traube and others have recorded such cases. Volmer had a case in which typhoid symptoms appeared first, then rigors, and the disease was recognized. The third and fourth groups comprise cases in which the temperature curve corresponds more or less closely to that of intermittent fever, with periods of paroxysmal exacerbation alternating with apyrexia, sometimes for a short period, regularly quotidian or tertian in type. In the third are those cases in which the signs of heart disease are ill-marked and null till near the close of life, and in the fourth those in which heart disease has been long established, the intermittent fever occurring as a final complication. He details four such cases. This disease was not known to physicians of the last century, Morgagni, for example.

With regard to the treatment, if endocarditis occur in connection with rheumatism, as it often does, the treatment for rheumatism is the best treatment for the endocarditis,

---

* *Medical News*, June 24, 1882.

as it is for pericarditis. If you can recognize the disease as occurring alone, then I would advise the abstraction of a little blood; unload the vessels; relieve the valves of a part of their work; not, however, more than eight or ten ounces. You will be tempted to apply counter irritation over the pericardial region. Well, that is orthodox, but it is useless. There is no way of affecting, by counter irritation, the circulation in the endocardium. It is entirely dependent upon the coronary arteries of the heart, and these are given off at a point that you cannot reach to influence the amount of blood circulating in them by any outside application. But, if it will satisfy you better, if your conscience tells you the patient ought to have a blister, put on one; though it will prevent your listening so satisfactorily, it will do no harm, and I don't believe it will do any good. In quieting the heart's irritation a moderate dose of opium is of value. The more quietly the heart can do its work, the fewer the embarrassments, the less the mischief that will follow. The strain upon the valves, whenever they are brought into exercise, is considerable—for example, that against the mitral when the heart is contracted; that against the aortic valves after the heart is contracted, and while the arteries are contracted. So that if the heart does its work in the quietest possible way the mischief to these valves will be the least that you can expect. Then you have really only two things that are of much consequence, and these two are the abstraction of blood and quieting the heart with opium or some other anæsthetic.

There is another form of endocarditis which has been lately described, or within a few years, at any rate, that you will not be able to recognize at all. Still, it is well enough that you should know that such a disease exists. It has been called ulcerative endocarditis. It seems to be a sort of diphtheritic inflammation. A diphtheritic inflammation of a mucous membrane, of the trachea, for example, or the fauces, differs from a common croup or other membranous inflammatory action in the fact that the deposition that is the result of inflammatory action is to a certain extent incorporated with the membrane; so to speak, infiltrated into it. A croupous membrane can form and can exfoliate and leave the membrane entirely. A diphtheritic membrane can exfoliate, but it is not likely to leave the membrane perfectly sound and normal. Sometimes it does, but it is the tendency of diphtheritic inflammation to deposit the products of inflammatory action in the tissue. Well, that seems to occur in the endocardium, and it may occur in any part of the endocardium, not alone in the valves. It does not occur so frequently in the valves as on the body of the endocardium. I should be disposed to regard it as a sort of diphtheritic inflammation. The new deposit looks like a membranous patch. It is in patches. It does not extend over the whole surface, and when you examine it post mortem, you will think you can wipe it off with your scalpel, or with a rag, but when you attempt to do so it holds fast, and if you with your scalpel attempt to remove it you remove the membrane with it. This proves to be ulcerative in its results. It is not ulcerative in the beginning. It is simply a deposit of diphtheritic membrane on and in the endocardium. But it has not the power to perpetuate its life, if it has any, and in a little while it will begin to break down into fine granular matter, and perhaps some oily products, and as it does this it destroys the membrane in which it is deposited. The endocardium, then, is susceptible of destruction in limited portions of its extent in this way. As the inflammatory exudation breaks up and disappears it carries with it a portion of the endocardium. If that occurs upon the valves, as it sometimes does, it is very likely to weaken that half of the valve, that layer or fold of the valve, so much that aneurism may take place in this valve. That is, the remaining membrane, the membrane that forms the other half of the valve, will not bear the strain that is brought upon it by the regular course of the circulation, and it pouches; it makes a bag, or, in other words, an aneurism, and it has been known to rupture under these circumstances. Then, again, the ulcerative process may not be confined to the endocardium if it occurs upon the body of the heart. It may penetrate a little distance into the muscular layers and weaken them, and hence, also, aneurism of the heart has occurred, or rupture of the heart. This, as I have told you, is a form of inflammation, or a product of inflammation that you cannot recognize during life. It produces no physical signs. It produces no peculiar rational signs. It may be found hereafter that it is confined to those who have some diphtheria in other parts of the body. And at this point I may as well say to you that when diphtheria made its appearance in New York for the first time as an affection of the throat and air passages, I was interested to know what sort of thing it was, and I early procured specimens of the membrane for microscopical examination. I found that it was made of granules in great degree—granules that were held together by a pretty firm matrix, and that the matrix was transparent; that on the outside of the membrane, that is on the side exposed to the air, there were the common microscopic vegetations, two or three forms of them. And that in this respect the diphtheritic membranes differed from the true croupous membranes, being constituted in very considerable part of granules, and in part also of extraneous vegetations. At a later day it was claimed that these granules were a particular vegetation of themselves, and they have received a name — micrococci.

Micrococcus means a little berry. These bodies are rounded, and in these membranes are packed together pretty closely. As you read the journals and new books you see a good deal of micrococcus. It seems to have found its way into parts of the body that have been supposed before to be sealed to any outside thing. Well, they tell us that ulcerative endocarditis is a consequence of the deposit of micrococci. They are all little bodies, not more than the one twenty-five thousandth of an inch in diameter. They are very minute, therefore, and it is claimed that this deposition in the endocardium is made up in very considerable degree of these little microscopic bodies. That would only bring it into closer affinity with diphtheria if it prove to be true. But even if it is true I do not believe that the micrococci are micrococci at all. I think that they are a particular form of deposit of lymph in granular shape, and certainly there is no proof that they are vegetations. It is a mere statement of opinion by somebody, and that opinion has, like a wave, spread pretty far. I am willing to admit that the subject is under judgment, is under arbitration, and it may prove that these bodies are little vegetations. It is said one divides and makes of itself two, and the two divide and make of themselves four, and the four divide and make of themselves eight, and as this division goes on pretty rapidly, you will soon get high numbers. I do not deny the possibility of this being a vegetation, but the probability of its being a vegetation is very doubtful in my mind. Still, that sort of thing does occur in the endocardium; a something that has a granular form is deposited, and afterward leads to ulceration.

LECTURE IV.

MYOCARDITIS.

Now, a few words about myocarditis; and the first word I have to say is, that it is tautology. We duplicate upon muscle. Still, it is a term that has become current, and we must meet it as it stands. The term means an inflammation of the muscular structure of the heart, and I am a good deal in doubt whether it is really an inflammation.

There you observe, in this upper figure, a very marked change in the walls of the left ventricle. Myocarditis produces at first a little reddening where it is located, and afterwards it comes down to about this color. The disease is local; it does not affect the whole heart at the same time. It may be a spot that you can cover with the end of your finger—at any rate with the palm of your hand —rarely larger. Here is a heroic sort of thing. I ought to have shown it to you before. It is a heart ruptured where once fatty degeneration occurred. Upon the exterior here is a spot of myocarditis which has passed its active stage, and a material that I shall describe to you pretty soon has been discharged from the pericardium. If you lift this up you will see there is a hole through the wall, and this is in the right side—the effects of myocarditis, but not myocarditis itself.

The reason why I doubt whether the change that I am going to describe to you is the result of inflammation is that, so far as is known, there are no general inflammatory symptoms ; no particular increase in the frequency of the pulse ; no known elevation of the temperature of the body.

MYOCARDITIS.

It comes on quietly, while the patient is attending to his business, and nothing is known of it really, except by conjecture. It is occasionally found to follow pericarditis and endocarditis. It affects, perhaps, one half the thickness of the muscular structure of the heart, and may occur in the septum ventriculorum, or it may occur upon the exterior face of the heart, either upon the left or upon the right side. As it makes its progress the material that is effused is not pus. Examined under the microscope it seems to be made up of a transparent fluid and broken-down portions of the heart tissue. It seems to be rather a breaking up of the muscles of the heart, and we can conceive that it may be by obstruction to the circulation of the blood to such particular part. When the disease is matured you find what you would call a little abscess, and what has always been called heretofore a little abscess; but the material, as I stated to you, that constitutes this apparent pus is not made up of pus corpuscles, but débris. This material is so fine you can only recognize its quality by the microscope. It at length is discharged, either outwardly into the pericardium, or inwardly into the cavity that it may be affecting. If the latter accident occur, it may do mischief in the way of producing embolism in distant parts of the system. More frequently it is evacuated exterior to the heart and within the pericardium. And in this latter case it seems to undergo absorption and is carried away, no remains of it being left. But after this evacuation of the débris of the muscular tissue, and to a certain extent of this connective tissue, there is a chance for cicatrization. The connective tissue which covers the muscular fibres is of itself sufficient to start a healing process, and if the heart is not overstrained there may be a cicatrix formed over the depression; a cicatrix of fibrous tissue, which, of itself, gives strength to the weakened part of the heart. It may rupture before this cicatrix is formed upon it. When the heart has been in that way affected, and the cicatrization has taken place, you may examine it at post-mortem, and find that it has not ruptured—the cicatrix has been a protection against that; but you find a depression covered by a whitish connective tissue, which is sometimes of very considerable strength. This is, in substance, myocarditis. And you can see now how myocarditis, in the particular specimen I gave you last, may produce a weakening of the muscular structure, so that it can be ruptured. That, together with the effects of the débris, this mixed matter, upon the circulation constitutes the whole of the bad history of myocarditis. There is a certain amount of inflammation of the heart itself when there is endocarditis and pericarditis, as I have already told you. The most common result of that, however, is simply an infiltration, into the muscular tissue, of some serum; a sort of œdema of the heart, which readily goes away after the inflammation has passed off. It is only in rare instances that this myocarditis occurs in the same relation; that is, in connection with endocarditis and pericarditis.

It occasionally happens that a part of the wall of the heart will yield to the contraction, or rather to the pressure of the blood upon it, and pouch out. You have a plate illustrating that now in your hands. That however, was a thinning of an already thinned heart; a particular part was more thinned than the rest, and yielded, so as to allow a pouch to be formed of the blood within the cavity. More commonly these aneurisms—for they are so called—occur in the inferior and right portion of the left ventricle. It becomes pouched out toward the median line and downward. This occurs sometimes to a very considerable extent. And it will form clots, layers of coagulated blood, as they will form in an aneurism of the popliteal space, or in the aorta, or in any other part of the body. As the blood coagulates in this expanded portion, it makes a covering which gives a certain degree of strength to this weakened part of the heart. After this has continued for a considerable time, a something of a white color, thick and resisting, of the character of a cicatrix, forms upon it on the inside. But this is not a cicatrix, because there has been no ulcerative process. But here was a weakened portion of the heart, and nature seems to know how to fortify it, and on the endocardium there is thrown out a certain amount of lymph, which becomes organized, and forms a new membrane of the connective tissue variety. You do not see that in this plate. You see only the layers that have been separated for the purpose of illustration ; layers of coagulated blood that have been imposed and superimposed upon this weakened part of the heart, and they are torn off in such a way as not to show you exactly the relation to the heart itself.

We do not know about this form of disease during life. It is found almost always after death. We can find an enlarged heart, and from the extension of the enlargement in a particular direction, we can infer that it is of the left ventricle, but we can hardly go beyond that. A distinct recognition, before death, of aneurism of the heart as it usually presents itself is exceedingly rare. I had lately—within a year, I think—a very remarkable case of aneurism of the heart, expanding in an anterior direction, and pushing its way through the intercostal spaces between the fourth and fifth ribs. I am not quite sure that is the point. There was a very decided pulsation. It was a case in Dr. Cole's practice, and he called me in to see it, and I could not make anything out of it but aneurism of the heart. It could not have been aneurism of the aorta, because it was below the point that that aneurism occupies. It pushed the ribs forward a little, and formed a cone in the intercostal space, like that upon the heart already shown you. Dr. Cole watched this patient for some months, and he at length died suddenly, as persons with aneurism are apt to die; and Dr. Cole asked for a post-mortem. He could not get it. The friends would not consent to it. He went to the coroner, and the coroner came and inspected the body, and he said, " Oh, yes, this is aneurism. We know aneurism well enough. A post-mortem is not necessary." And then Dr. Cole resorted to the priest, and tried to induce the priest to persuade them to let such a curious thing be seen, but the priest, even, hadn't the power to persuade them, and the mystery was buried in the grave. Still, I can hardly entertain any doubt that it was of this kind. Cases of the kind, you may understand, are rare; still in the enumeration of diseases of the heart it seems important to say a few words, even of the occurrence of a rare one.

Dr. Carl Huber[*] gives a number of cases of chronic myocarditis and disease of the coronary arteries, in which death occured suddenly, or with only preceding constriction or pain in the chest; and in which the cause of death appeared to be sclerosis of the coronary arteries and subsequent chronic myocarditis. The consequences of this myocarditis he thinks are aneurism of the heart, thrombosis, dilatation and hypertrophy. The symptoms are angina pectoris, stenocardia, and asthma, 'occurring in paroxysms usually after excitement, physical or mental, sometimes after dinner. In some there is irregularity or intermittance of the pulse, sometimes a sudden giddiness with temporary loss of consciousness on stooping, walking quickly, or going upstairs; sometimes there is collapse. It has nothing to do with endocardial or pericardial disease, but on arterial sclerosis. The general cause is alcoholism or syphilis.

---

[*] *Med. Rec.* July 28, 1883.

Dr. Beverly Robinson * read a paper before the Practitioners' Society, March 2d, 1883, in which he detailed some cases in which the death was sudden or came rapidly in the course of typhoid fever and diphtheria, and one or two in which recovery took place when symptoms had appeared, which in other similar cases had ended in death. This occurred twice in one patient, and in others once or twice before the fatal sinking. In all these it is stated that nothing abnormal could be discovered by examination of the heart. He thinks there is often under these circumstances a sudden and considerable dilatation of the cardiac cavities, and especially of the right heart, and he says, "Coagula may then form, or if the heart be immediately and strongly stimulated, the imminent stage may be tided over."

Dr. Kinnicutt said that in various acute infectious diseases, in diseases accompanied by high temperature, as well as in pericarditis and endocarditis, pathological research had demonstrated the very frequent existence of a parenchymatous myocarditis. The inflammation may be diffused or circumscribed. The cardiac muscle became swollen and pale, numerous minute granules partly soluble in acetic acid made their appearance in the muscular fibre, the transverse striae become indistinct, and finally disappear. This process may proceed to a condition of true fatty degeneration. Might not these changes account for some of these sudden heart failures?

Dr. C. S. Dana said that French observers had recognized the fact that in acute infectious diseases the poison sometimes especially attacks the heart. In typhoid fever a "cardiac form" has been described. Sudden deaths occur also in nephritis and pneumonia.

* The *Medical Record*, May 5, 1883.

LECTURE V.

HYPERTROPHY OF THE HEART.

WHEN I was a student of medicine hypertrophy was described under three divisions: First, simple hypertrophy; second, eccentric hypertrophy; and third, concentric hypertrophy. We do not use those terms now. It is hypertrophy, and hypertrophy with dilatation. Hypertrophy with contraction, concentric hypertrophy, is probably, in every instance, no hypertrophy at all. The last act of the heart was to contract, and to force out the blood that was, just the instant before, within its cavity. It is very much contracted, its cavity is nearly obliterated and it may be taken for a diseased condition. But it is a natural office of the heart to contract, and if death strikes at the instant that the contraction is completed you will, of course, find no blood in the ventricles of the heart: you will find it smaller than natural, and that would not be consistent with the term hypertrophy. The walls are thicker because they are contracted. I think it would be found that every instance that was, of old, called concentric hypertrophy could be resolved into the condition that I have described. So that now we have hypertrophy, and hypertrophy with dilatation.

The first question that presents itself is, what is hypertrophy of the heart? Well, the obvious answer is, enlargement of its walls; increase in the bulk of matter that constitutes the walls of the heart. Well, what constitutes the walls of the heart? Mainly, muscular fibres, and some connective tissue. Then it is the muscular fibres of the heart that are increased; and how increased? I asked myself that question a great many years ago, and I resorted to some old specimens I found at the hospital for the answer. A point to be investigated was, the size of the muscular fibre in hypertrophy of the heart; is it larger than the muscular fibre in the heart of natural size? I procured about twenty or twenty-five, and drew fibres from different parts of the heart, from one ventricle, and then from the other, and compared those from the natural heart with those from the hypertrophied heart, and though occasionally I would find enlarged fibre the average was pretty nearly the same in the healthy and in the hypertrophied heart. Well, then, the conclusion is inevitable that if the same proportion exists between the size of the muscular fibre of the natural and of the hypertrophied heart, there must be a multiplication of the fibres of the muscles in the latter, and that is a fact. There is a new production of muscular fibres. From what origin they spring I cannot tell you, but nature has so arranged it that if the heart needs additional strength it can get additional fibres.

The size to which the heart may grow is, perhaps, limited to sixty ounces. When we meet next I intend to show you one that weighed fifty-seven ounces—within three ounces of the largest heart the weight of which has been recorded. The natural weight of a heart in a full grown person, not remarkably broad upon the shoulders, is eleven ounces; if, then, it is increased to sixteen or eighteen ounces, you see there is some hypertrophy; if it is increased to twenty-five or thirty, there is very considerable hypertrophy; and if it is increased to anywhere in the neighborhood of sixty ounces it is *cors bovinum*, an oxheart, and here I may say that that enormous size is never attained except in persons in whom disease of the heart began in infancy. This, the largest heart I possess, perhaps the largest heart in the country, was taken from a man who, at six years of age, was taken over by his physician from Randolph, Vermont, to Dr. Nathan Smith, and was shown by him as an instance of heart disease arising from inflammatory fever, rheumatism. The boy had had articular rheumatism, and his heart became involved, and Dr. Nathan Smith had interest in it as an instance showing there might be some connection between articular rheumatism and disease of the heart. It was fourteen years after that—more than that—twenty odd years, before the connection between articular rheumatism, pericariditis, and endocarditis was made out; so that, really, had it been published, that discovery of Dr. Smith would have been put first, as showing the relation between rheumatism and inflammatory affections of the heart. The boy grew up, attained the age of twenty-eight years, and to show you his was not a useless heart, he was foreman of a manufactory in the immediate neighhood of Springfield, Mass. He was only incapacitated for work about a month. He kept his men all in line, all at work, until his breath became so short, probably through involvement of the kidneys, that it was so painful to him he was forced to give it up. He remained at home after that, and died about a month after ceasing to work, and yet carried this enormous heart. I have never seen a very large heart in a person in whom it did not begin to grow in early life.

The heart may be hypertrophied in the wall of one of its cavities. The hypertrophy usually found about the heart is that of the walls of the left ventricle. The left ventricle is hypertrophied oftener than any other portion of the heart. You will find, also, hearts that are hypertrophied in all of the cavities; perhaps some such among those that I shall show you at our next meeting; and others, perhaps, that are hypertrophied in the two ventricles; and one or two, I think I have, that are hypertrophied in the right ventricle, the right auricle, and the left ventricle. So that hypertrophy may be single, or it may occupy each one of the walls of the heart cavities.

The causes of hypertrophy are various. The most common cause is the need of more strength in the walls of the cavity that is immediately affected. Obstruction, then, at the aortic opening, as has already been stated to you, is attended by hypertrophy; but the heart does not seem to know exactly when to stop growing, though it does know when to begin to grow, and it may go on to become diseased. A moderate hypertrophy, then, with obstruction at the aortic opening, may not be regarded as a grave disease. Then, regurgitations will lead to hypertrophy. That the obstruction not requiring increased strength in the walls of a cavity will not produce hypertrophy, you will see in the fact that stenosis of the mitral valve is attended by no change in the left ventricle. If there is regurgitation, then the left ventricle may become hypertrophied. It needs more strength for the work it has to do. But if it is simply an obstruction of the blood coming from the left auricle into the left ventricle, and the left ventricle becomes in the end fairly filled, and does not throw the blood back again into the auricle, then there is never any hypertrophy of the left ventricle. It is only in regurgitation with this lesion that hypertrophy comes. Here, then, is one cause.

Another cause is increased strain upon the heart; an increased demand continued for a long time, attended by a nutritive, non-deteriorated condition of the blood. You see why I make that limitation; because a man with typhoid fever may have his heart beating rapidly and pretty strongly for four weeks continuously, and no hypertrophy follow. But then his blood is deteriorated; all his tissues are, throughout. If by mental or physical excitement his heart should beat this way, at the same rate and for the same length of time, and his blood were in a nourished condition, his heart would become hypertrophied to a certain extent. Disease that produces increased action of the heart and full action of the heart, does not, then, cause hypertrophy, unless the blood is in good, full, nutritive condition. Occupations that are attended by great mental excitement and by heart beating are dangerous in this respect, producing hypertrophy without valvular disease. The little tyrants that command on vessels at sea sometimes are very passionate men. I have seen them beat their men about and swear at them as if they belonged to our army of Flanders; and those men are very apt to get hypertrophy of the heart. They do not observe the rule of Scripture, let not the sun go down upon thy wrath. They are wrathful at night. A French physiologist has interested himself in the cardiac condition of the domestic fowl, and he found that where one rooster had to be the husband of ten or twelve hens, he was apt to get hypertrophy of the heart. This may be a warning in regard to sexual indulgence.

Then, again, hypertrophy comes from causes that are not known. We say in general that it is an error of nutrition, but why, we do not know. Persons of quiet temper and of easy life, not subject to particular excitements, and having no valvular disease of the heart, do once in a while get hypertrophy of the heart; but it is uncommon; decidedly uncommon.

The time seems to be passed when it was necessary to ascertain the effect of the contracted kidney on the heart by experiment on animals. There appears to be very general assent to the proposition that it causes hypertrophy of the left ventricle. But in the reports of those who have studied the matter experimentally there is a remarkable disagreement, and it is worthy of notice that while the clinical studies have been settling the question the experimental should be so inharmonious.

Dr. Straus, of Paris, tied one ureter in twenty guinea pigs to cause atrophy of the kidney to which it belonged. These animals were killed four to six months after this was done, and the left ventricle was found enlarged in every one of them. The hypertrophy was not great— only $\frac{1}{6}$ to $\frac{1}{4}$ the original weight, but the time was short. The muscular fibres were healthy.

Grawitz and Israel asserted that while hypertrophy of the heart might follow a kidney lesion in old animals in which the other kidney did not grow to compensation this result was not to be obtained in young animals. But nearly all of Dr. Straus' animals were young.

Rollet [*] records the following case of hypertrophy confined to the left ventricle, the walls of which were pale and presented here and there small spots of sclerosed tissue. The aortic orifice was normal. A fibrous band was found extending from the under surface of the aortic lip of the mitral valve to the intraventricular septum and aortic portion of the ventricle at a little distance from the sigmoid valves. The new growth narrowed the aortic cone to such an extent that one finger could with difficulty be introduced. The author thinks this band was produced by endocarditis before birth. A similar case affecting the right ventricle is recorded by Dettrich, 1849.

The patient was a woman forty-seven years of age, admitted with palpitations, thoracic pain, dyspnœa and cephalalgia; cardiac rhythm irregular. There was an increased area of dulness, apex beat strong and to the left of its normal place. There was a distinct thrill, perceptible only under the lower part of the sternum, to the left of that bone and at the apex. There was a rather pro-

---

[*] The *Medical Record*, Oct. 7, 1882.

longed systolic murmur of greatest intensity in the fourth interspace at the left border of the sternum. The second sound was audible at the base and in the carotid, showing the integrity of the sigmoid valves.

Now the question comes, how to recognize hypertrophy of the heart. A heart like that one that is passing around, which I told you is hypertrophied, occupied a greater space in the chest than it did when it was of normal size, and what will that do? It will crowd the lung away. The space over which a heart of natural size will practically come in contact with the walls of the chest, or, in other words, not be covered by lung, is about equal to two superficial inches; but when a heart grows, as that has, it must crowd the lung out of position. In a healthy person you can hear the respiratory murmur over the whole præcordial region; but when you listen to a person who has considerable hypertrophy of the heart you will very likely fail to find the respiratory murmur over the anterior portion of the heart, because the lung has been crowded away. That comes to be, then, one sign for the recognition of it. The same thing, however, occurs when the pericardium is dilated with fluid. The lung then is necessarily crowded away in the same manner.

Another point is that the region of dulness is increased more or less markedly. The point to which the apex of the heart can be pushed to the left by its hypertrophy is perhaps six inches and a half from the median line. Then you get to about the outer curve, the outer limit of the curve of the chest. It may be pressed backward beyond that, but I do not remember to have measured a heart which extended to the left more than six inches and a half. I have found it six inches a great many times, and five inches and a half a great many times. You can measure it in either one of two ways. Find the apex, and be sure that there is no heart beat beyond that, and you can pretty safely measure up to that point from the median line. But it is always better to make percussion, so that you will be sure of the point to which you should measure. Dulness over the lung upon an enlarged heart is very decided. The line of division between them is pretty clearly cut. Yet, even suppose you get dulness over six inches from the median line, your diagnosis is not yet made. The same thing may happen with pericarditis and an enlarged pericardium. Then seek for the apex of the heart. If there is marked enlargement of the right side of the heart the apex will be tilted up toward the axilla. It will be, perhaps, as high as the fifth rib, following the curve of the rib upward, or it may not be as high as that. Its direction is upward and outward, and to the left. If it is the left side of the heart exclusively that is hypertrophied, the apex may be situated in the seventh intercostal space, outward, downward, and to the left. Of course it will be in the sixth before it is in the seventh. It is in the intercostal space, of course, that you can most easily feel the impulse of the heart. And then, again, you must make percussion to verify the indications of the apex-beat. There is nothing that is likely to trouble you in the attempt at diagnosis but pericarditis. It is true, the heart is sometimes the seat of tumors; and the pericardium becomes enlarged sometimes by air, but that will not give dulness. In general practice, with the exception of the extraordinary cases, that you do not expect to see, these two things will give you the only trouble there will be. Hypertrophy is otherwise easily recognized. There is a point, however, to be borne in mind. I examined a gentleman yesterday who had been educated to the profession, but had gone into business. He had pleurisy filling the left side of the chest, and his heart was pushed over so that the apex beat was about in the position on the right side that it would naturally occupy on the left, crowded over by this accumulation of fluid in the left side of the chest. Now, a person who has hypertrophy of the heart may have pleurisy as well as anybody else, and the heart may be displaced in the same way. It may be displaced to the left also. An effusion upon the right side will encroach upon the right side, will encroach upon the mediastinum sufficiently to crowd the heart over an inch further to the left than it naturally would be. Such a case as that you may mistake for hypertrophy when it is only a displacement. It will be important to know, then, whether there is any effusion upon the right side of the chest.

Then, again, the history of hypertrophy and of pericarditis will be found different. Hypertrophy is a chronic affection, and a man who has it, and is sent to you for it, will have had for a good while some shortness of breath in going up stairs. Then, again, the dulness extends, in pericarditis, as I have already told you, in a direction different from the dulness in hypertrophy. You get a certain degree of dulness in pericarditis, with copious effusion, on the right side of the sternum, and you get it above the third rib. It is only effusion in the pericardium that does that. The physical signs of hypertrophy, therefore, are pretty clearly made, pretty clearly cut, and you need hardly make any mistake about it.

The rational signs are more obscure. You cannot be sure, by any study of the general symptoms, that you have hypertrophy and nothing else. You can pretty easily determine, from the rational symptoms, that there is heart disease, but exactly what it is, is almost impossible to determine. The usual symptoms of heart disease are to be observed in these patients—that is, a certain degree of shortness of the breath on exertion, frequently a disposition to rest the head on two or three pillows, in-

HYPERTROPHY OF THE HEART.   131

stead of one, in sleep; a disinclination to sleep upon the right side of the body, not because it gives them pain, but because it disturbs them, it makes them nervous; they feel the heart beat, and that makes them nervous. Palpitation cannot be relied on as a means to distinguish hypertrophy. It is wonderful how much of thumping the heart will have, and the patient be entirely unaware of it in some instance; and it is really surprising how little of real palpitation will give other persons much uneasiness. There is a real palpitation and there is a subjective palpitation. Persons of particular habits will become nervous and feel palpitation when there is only natural action of the heart. It arises alone from their increased sensitiveness or the heightened state of nervous perception. And then, as I said before, a man who has real hypertrophy of the heart, who has been accustomed to hard knocks in life, may pay no attention to it, while it may lift your head half an inch at each beat when you apply your head to the chest. It knocks like a hammer against the face and ear; and yet, ask him if his heart palpitates. "O no, O no, my heart is good!" But palpitation is not a certain guide to diagnosis. The man we saw at the clinic yesterday had no palpitation; that is, nothing that you could feel with the hand. You could feel that the heart was beating, but not with extraordinary force. He had had his disease for a considerable time, and it is to be presumed that the heart had lost some of its energy. But even in persons in whom the disease may be gradually growing there may be decided palpitation, and there may not be, so that you will have to consider what is the cause of the palpitation, if you find it.

Shortness of breath is common to almost all the forms of cardiac disease, so that there is nothing particularly distinctive in that. Then comes the question, what can be done for the relief of these patients? and then, again, as all the diseases, or nearly all the diseases, of the heart, have about the same physical history, their treatment will be substantially the same, so that I will say what is to be said about that further on.

LECTURE VI.

DILATATION OF THE HEART.

THERE is another condition of the heart that deserves particular attention. I say particular, as it is not nearly so common as hypertrophy with dilatation. It is dilatation alone, and this brings me to the question, what produces dilatation of the heart, whether it occurs with hypertrophy or singly? The answer to this question, I think, has been given by Niemeyer: dilatation of the heart is caused by a reflux of blood into it, which blood has just been thrown out of a cavity. For example, the blood is thrown into the aorta, there is a reflowing, a reflux of it into the ventricle; perhaps one half of it that was thrown out comes back again, and this is in the period of non-resistance of the heart; its muscle is all relaxed, flabby. If it were to come during contraction there would be resistance to enlargement, but the muscle is, so to speak, dead at the moment; it is in the period of repose that this reflux takes place. It stretches it, together with the blood flowing in from the auricle. It is compelled to contain more than it was made to contain, and the result is a moderate dilatation, and this continued seventy-two times a minute, for days, and weeks, and months, will produce an effect. One occurrence, probably, would not be felt, but a repetition is what produces the dilatation. Well, this dilatation, then, is caused by an unnatural flow of blood into the ventricle, and the heart, feeling its contents, naturally acts with more than ordinary force, so that hypertrophy and dilatation often go together, step by step.

Dr. Peabody * showed to the N. Y. Pathological Society a heart of which the following is a description: The endocardium was generally thickened, as were the aortic cusps. The chief lesion was in the muscle of the heart. It had been replaced by connective tissue largely in two places; the most marked was a circular spot of nearly two inches in diameter between the papillary muscle and the attachment of the aortic valve, and there was distinct aneurismal bulging of the septum ventriculorum and also of the left border of the left ventricle. The depth of the sac was three quarters of an inch. A second sac was near the apex of the ventricle in the septum. The muscular tissue was in the state of brown atrophy. The coronary arteries open at their origin; their branches were obstructed or occluded by a growth from their lining membrane. Patient was a man fifty-five years old.

I told you, in speaking of hypertrophy, that I would show you a heart that is, probably, the largest that has been noticed in this country. This is the specimen. There are several interesting things about this: first, that it is, probably, the largest heart that has been in this land; second, its history is well preserved. I gave you the history in a previous lecture. The patient could carry this large heart only because he grew up with it. These large hearts, as far as I have noticed, always begin to be diseased in early childhood, possibly in infancy, and the person becomes accustomed to the extraordinary develop-

---

* *Medical Record*, July 29, 1882.

ment, and so it gives little inconvenience. It grows with his growth, and may attain this monstrous size.

I told you, when speaking of pericarditis, that I would exhibit to you a heart to which the pericardium was attached. You observe here is the pericardium, and it is attached to the heart. The patient had had endocarditis and pericarditis, and this is an instance in which we can pretty safely say we have twenty-two years from development of the pericarditis till death. You will see the manner in which the pericardium is trying to separate itself from the heart. By just raising up the flap of this pericardium on either side you will see threads running from one side to the other. They have been considerably torn up by handling, but still I think you will notice them better on the upper side. This pericardium, had the man lived long enough, would have been separate. The attachment, as you see, is loose. Then, as a specimen of aortic disease, it is extraordinary. Opening the aorta by "its ears," and getting down to the valves, you see they are thick and leathery. They are more than a line in thickness, nearly an eighth of an inch in thickness, and they are much distorted, and are, you see, quite incapable of performing their office. They feel hard, like sole leather. Then the right auricle, together with the right ventricle, you observe, is somewhat dilated, but very much encroached upon by the septum ventriculorum. The mitral valve is not so very much diseased. That is to say, its offices are not very much hindered. It is thick, scarcely leathery, its posterior valve is contracted. It is contracted considerably, and I am not sure that it would perform its office perfectly. But the amount of disease there is trifling compared with that which effects the aortic valves.

Here, too, has been pericarditis. You observe how rough the heart is. There is something of that thready appearance, not so much, however, as in the other specimen, in this pericardium. I do not know the history of this case as I do that, but you observe here is very marked shortening of the valve. That valve is not one half its proper height, and it would be impossible for its parts to meet in the middle of the artery and prevent regurgitation. In this instance the right heart is more hypertrophied than the left—that is, in proportion to its natural thickness the walls are thicker, and in regard to capacity it seems to be just about as great as the left. The thickening here is quite remarkable. The tricuspid valve does not seem to be diseased. The aortic valve is firm, and shortened to a certain extent, and here again the posterior more than the anterior curtain of the valve. But the great defects, or rather the principal diseases of the heart, are the hypertrophy, particularly of the right side, and the valvular disease at the aortic opening. The pulmonary valves are about natural.

Here is another specimen of marked hypertrophy, and here again you observe a septum of the aortic valve thickened and drawn down. The valves, you observe, are very imperfect, and allowed regurgitation, while the mitral valve is in tolerably good condition. The right ventricle is small, comparatively. It will not hold one half, it will hardly hold a quarter of what the left will, and that of itself is a kind of disease. The blood can be thrown out from the left heart a great deal more rapidly than it can be received by the right; and the result is exactly the same as when the valves are defective.

I was speaking to you, when we were last together, of dilatation. I had begun to describe to you the peculiarities of that disorder of the heart. We will go on with that, and I will show you at our next meeting a specimen that is rather striking; a dilated heart, in which, if I remember right, all the cavities are dilated, and they are stuffed with hair, so that the heart has about the size that it had in life, and yet in the cuts made into the walls you will see to what a very marked thinness it has been reduced.

Dilatation, I told you, considering it in the sense in which we are using the word now, is a thinning of the walls of the heart. We have hypertrophy with dilatation, in which there may be considerable thickening of the walls of the heart, but this is spoken of by a distinctive name. We now refer to simple dilatation of the heart, in which the walls of the heart of the left side come to be little more than a line in thickness. The walls of the right heart are from three to five lines in thickness in the normal state, that is, three at about half an inch from the base, and growing a little thicker toward the apex, and at the apex five lines in thickness. But this heart that I shall exhibit to you at our next meeting will give less than a line in thickness of any portion of the left ventricular wall.

The first question that presents itself to us is, how can this thinning occur? A good many speculations have been offered in explanation of this occurrence, but I am inclined to think that the German explanation is the best: that it may occur in two different ways: First, by a reflux of blood when the blood is in a non-nutritive state, dilating the natural walls of the heart till they are thinned down to a condition I shall exhibit to you; the other, that hypertrophy with dilatation may have occurred, and that by fatty degeneration, which I shall explain to you a little further on, the material constituting the muscle of the heart may be absorbed, taken up, carried away, leaving very little but the two serous membranes, the endocardium, and the pericardium. That implies, of course, very marked feebleness in the heart's action. Simple dilatation of the heart to the extent of becoming a disease is a rare occurrence. I have met with two or three cases, and that is all.

The indications of it, the means of recognizing it during life, are both physical and rational. Among the physical signs will be one that belongs to hypertrophy, extended region of dulness. The lungs will be pushed away, in the same way, from their natural position, and in auscultation for the respiration you will find it silent over the dilated heart. That is, the lung is pushed so far away that the sounds of respiration will not be brought to the ear placed in the middle portion of the pericardial region. There is no palpitation proper. It frequently happens, however, that the patient complains of palpitation. His sensitiveness is increased at the same time that the action of the heart is rendered feebler; though he complain of palpitation you put your hand over the heart, and you can scarcely feel it beat. In one case that I saw it was remarkable that the contraction of the heart was in a wave. It was not a sudden contraction of all the muscles of the heart at the same moment, but the contractile effort seemed to pass up under the hand. Beginning below I could feel the wave pass the length of my hand. That is, so far as it was applied to the intercostal spaces. I could not feel it, of course, through the ribs, but the intercostal spaces being thin, as I exerted some pressure they would indicate to my hand the kind of action that was taking place in the heart.

A feeble beat of the heart, then, with an increased region of dulness, and a consideration of the kind of sound, will be the chief physical signs. The sounds of the heart are changed in relation to each other; the time is changed, or the rhythm, as it is called. As the heart beats naturally, the second sound follows instantly upon the cessation of the first, and then there is a period of rest, about as long as the period occupied by the first and second sounds together. In this condition you are very apt to have the motion of the heart, or rather the sounds of the heart, resembling the ticking of a watch, the first and second sounds at even distances apart. The first sound is short, scarcely exceeding in duration the second. Tick, tack; tick, tack; in about that relation.

Then the appearance of the patient gives a marked indication of dilatation. The force with which the heart contracts upon the blood is so far diminished that a small quantity is sent into the arteries at any one beat, at any one time. This does not dilate the arteries, does not keep them of their natural size. They are apt to grow small therefore. I referred the other day to the filling of the ventricle during the period of rest of the heart. I perhaps should have said the ventricle has no power of dilating itself, but it is dilated by the force that it has impressed upon the blood when it went out from the ventricle into the arteries. The residual force of the circulation when the blood returns to the heart is really the dilating power. In that sense it dilates itself. The venous power is considerable, as when we tie the arm to bleed a man the force of the current is such as to throw the blood quite a considerable distance from the arm. This force is not all exhausted when it comes to the right heart, and it is that residual force impressed upon the circulation by the contraction of the left side that opens the right. In this case the force impressed upon the blood going out of the right side of the heart opens the left ventricle, or causes the left auricle to fill, and this is really the cause of the filling of the left ventricle. · Now, in this dilated heart, you have a very feeble force impressed upon the blood when it comes out of either ventricle. It will, of course, flow into the opposite ventricle in an equally feeble stream. It will dilate the veins, because there is not force enough behind to empty them, and you get œdema of the face, of the feet, and of the surface of the body, and you get at the same time enlarged veins; the blood seems to rest in the veins. Of course it does not absolutely rest, but the current is slow.

The oppression of the breathing is very marked in these cases. It belongs to almost every form of disease of the heart, but is developed earlier, and is, perhaps, more severe in this than in any other, and that depends upon the failing force with which the blood can be circulated through the lungs. There are at least three distinct causes of dyspnœa besides the nervous causes that would produce the same thing. One is an over-charged or congested lung or lungs; another, very slow circulation of the blood through the lungs; and the other, some physical obstacle to the dilatation of the lungs by the air entering them. These three conditions all produce dyspnœa more or less severe, and why? You would suppose that, mechanically, they would not act in the same way. Mechanically, they do not, but physiologically they do. They each one interfere with the aëration of the blood. When the lungs are congested the blood is circulating slowly through them, and though the blood that is in the lungs may become aërated it is not given to the system with sufficient rapidity to answer the demands, and consequently a sense of dyspnœa. And, then, too, when the quantity circulating through the lungs is very small and the current slow, there is not enough aëration of the bulk of the circulating blood, and there is dyspnœa. Then, when both pleuritic cavities are filled with serous fluid, or any other fluid, and the lung room is not great enough, there is not space for very much aëration of blood, and there is dyspnœa. And these three conditions of the lungs explain most of the dyspnœas that occur with the diseases of the heart, and in none of them is that dyspnœa more perceived, more apt to cause suffering, than in this dilatation of the heart. These are the leading facts by which you are able to recognize simple dilatation of the heart, if you should happen to meet it.

With reference to its treatment I will say a word. There is something to be said in regard to this that does not belong to all forms of diseases of the heart. I see recommended in some of the text-books, and in special papers on this affection, as a means of relief, bleeding—bleeding from the arm. Well, I will tell you my experience. When I was house physician in the New York Hospital a man came in in whom we recognized dilatation of the heart, and it was in that particular man that I got that wavy motion of the contractile fibres of the heart. My principal, seeing how much he suffered, said, bleed him. I looked at him with surprise, and said, "Doctor, I would not dare to bleed him. He cannot lose five ounces of blood without dying." "Pooh, pooh!" says he. "If you don't like to bleed him, give me the lancet." He tied up the arm and took the lancet. I was accessory in holding the bowl, and the vein was opened. The man bled a little spurt, it may have been four or five, possibly six ounces, and rolled over on the bed—I had him sit on the bed on purpose—rolled over on the bed, and breathed no more. Of course that did not make me very much in love with bleeding for dilatation of the heart. And yet, even in Niemeyer, if I remember right, it is recommended, but, I suppose, theoretically, inasmuch as bleeding does in some conditions of dyspnœa give relief. But I do not think it applicable to the dyspnœa which arises from dilatation of the heart. Well, then, there is very little you can do. The man cannot take any exercise. He would drop down of dyspnœa if he were to make any great exertion. He may walk about the ward a little, if he is in a hospital, or in his room, if at home; but that is about all he can do. He cannot go up stairs and can scarcely go down. About all you can do is to induce him to take as much food as he can digest, with the hope of enriching the blood a little, and delaying the progress of the disease; to give him chalybeates, which will help in this change of the condition of the blood, almost any form of iron that is a favorite with you; it matters very little, for they all go to about the same thing; and create an active condition of the kidneys to carry off the œdema. This latter is applicable to almost all the forms of cardiac disease where dropsy has occurred. I shall recur to the matter of diuretics by and by.

With these remarks on dilatation of the heart, we will leave that subject and turn to another condition of disease of this organ.

Prof. Maragliano* gives the results of the administration of strychnine in dilatation of the heart as follows: 1. In two or three days the size of the heart was reduced, and in five or six days very considerable dilatations were caused to disappear. 2. If immediately on the reduction in the size of the heart the strychnine were withdrawn, the dilatation was frequently reproduced. 3. The daily dose of sulphate of strychnine required was from $\frac{1}{32}$ to $\frac{1}{20}$ of a grain.

---

* The *Medical Record*, Jan. 27, 1883.

LECTURE VII.

FATTY DEGENERATION OF THE HEART.

You hear a great deal about fatty degeneration of the heart. You will hear a great deal more than you will see. I scarcely ever see a fat man advanced somewhat in years who has trouble with the heart, that the attending physician does not say, "Why, doctor, don't you think he has fatty degeneration of the heart?" My reply is uniform, "I cannot diagnosticate fatty degeneration of the heart, and I have no right to assume it in any case without proof." You will see where the difficulty lies pretty soon. Fatty degeneration, to the extent of producing real disease, is quite a rare occurrence; occurring, to be sure, now and then, but in comparing it with dilatation and hypertrophy together, it hardly bears the relation of one to fifty.

There are two kinds of fatty degeneration of this organ. One I have sometimes designated as adipose degeneration, or adipose deposit upon the heart. The other is an integral degeneration of the muscle of the heart into fat. The latter is called, commonly, Quain's degeneration, as he was the first to describe it. The two are very easily distinguished by the appearance after death. But neither of them can be confidently recognized during life. The latter, Quain's disease, you will be more likely to distinguish than the other. The fatty degeneration of the adipose variety is merely a deposit of fat upon the surface of the heart. I have some better specimens than this which I can show you, perhaps, at our next meeting, though this is well enough marked to give you an idea of what is meant by it. Always there is some fat upon the exterior of the heart in the pericardium, but you will observe in examining this specimen that here it has encroached somewhat upon the muscular element of the organ. The whole heart is covered with a more or less abundant layer of fatty tissue, but chiefly upon the right side. Well, that is the part of the heart that is most commonly overloaded with this fat. You take a portion of this fat and examine it under the microscope, and it will present exactly the same appearance as adipose tissue taken from any portion of the body—a multitude of cells filled with oil. It is a healthy fat enough, but it is out of place here, for the reason that as it becomes abundant the muscular heart becomes weakened. But many persons, no doubt, carry this condition for a considerable time without knowing it. The heart performs its office with fair regularity and with fair force. But in a few instances it will happen that the fatty encroachment is unequal, and it weakens one part of the heart more than another; as, for example, the septum ventriculorum may be just as strong as it ever was, while the wall opposite is weakened. The result is that the heart sometimes gives way first at that point; at the point where the greatest weakness of its muscular tissue is it bursts itself. I will refer to that again a little further along. This is the heart, then, and you may examine it and see what it means. The fat is very abundant upon the base, but that does not do so much harm as that deposited along the course of the ventricular wall. This heart is also hypertrophied.

I told you the heart beats with perfect regularity, and that it is not enlarged in any such degree as that you can recognize it by physical signs. Well, then, how can you recognize it at all? Well, just make up your minds, now, and forever, that you cannot recognize it, except after

144   FATTY DEGENERATION OF THE HEART.

death. And we will not spend a great deal of time in talking about fatty degeneration of the heart, when you cannot tell what is the matter with the patient when he has it.

Now, the other kind is altogether different in its effects upon the heart, and altogether different in its pathological aspect. The heart muscle is a striped muscle. It has cross markings, lines in the direction of its length, and lines across dividing those muscles, as the muscles of the arm and leg, and all the voluntary muscles of the body are divided into little checks. Well, these checks, when Quain's degeneration occurs, come to hold first a little oil globule. A muscular element seems to be removed from it. The place of the muscular element is taken by oil drops, minute, microscopic drops. I can show you a plate the next time that will, perhaps, give you a better idea than any of my drawings, but you may get some idea of it from this drawing. Well, then, this sort of heart has become useless. No, not quite, because this sort of degeneration is unequal in the walls of the cavities. It will be here marked, and at another place scarcely noticeable. The result of this production of oil within a muscle, covering the sarcolemma, is to change the color of the heart, and to change its consistency. Here, for example, are three specimens of this kind of degeneration. This one is not so very much softened. It seems to be small. The walls of the heart are thin, but the cavities are not dilated. The peculiar color that the oil gives to it is noticeably yellow. Here is another which is soft and yellow, and the walls of the heart are a little thinned, but not reduced so much as that you would be able to recognize it during life. And here are several other specimens which you will examine with regard to color, etc.

In regard to the recognition of this form of disease, it is more nearly possible than to recognize an adipose deposit upon the exterior of the heart. For example, I was called to see a lady, not quite middle aged, in whom there was marked irregularity in the action of the heart. I measured its size; that was natural. She was of an age that would hardly admit of the supposition that she had bony degeneration of the heart structure. That belongs to advanced age. She was not more than thirty, or perhaps thirty-five. She could lie down flat in bed—that is to say, she could lie down on her back without any pillow or bolster, and felt better in that position than with her head raised. She was dizzy, and disposed to faint when she walked about, and particularly when she made any exertion. The peculiar irregularity—I say peculiar; I do not know that that is the proper name—the marked irregularity, the irregular irregularity in her heart fixed on my mind that there was some disease of it. In using that expression, irregular irregularity, I ought, perhaps, to define it. When a heart beats on for five, six, or seven minutes, and then drops a beat, that is an irregularity; but there is a certain degree of regularity in its recurrence. You may count a certain pulse, and find that the sixth is dropped through a whole minute. But when there comes a kind of irregularity that you cannot describe, when the heart beats perhaps rapidly in this way and then holds up, beats slowly, and then directly begins with a sort of fluttering that is hardly describable in words, that is what I mean by irregular irregularity as contrasted with the regular recurrence of certain phenomena in the action of the heart. Well, this lady had that. The heart beat with an irregularity that is hard to describe. Now, it will be well for you to remember that that belongs to some particular disease of the heart; that that particular irregularity that you cannot well describe does not come of mere functional disturbance, but is always, as far as my memory goes, associated with a change of some kind in the structure of the heart or its valves. I said to the doctor who called me in, "I am confident there is some disease of the heart here, but I cannot tell what it is." Had I known the fact that in this kind of degeneration it is not uncommon that patients can lie down with the head low in bed, and feel better than when the head is raised, I might perhaps have made a reasonable conjecture. But at that time I did not know it—thirty or forty years ago. This lady died, and we had a post-mortem examination, and the heart I think is one of these. I examined various portions of it under the microscope, and found the kind of degeneration that I have been trying to describe to you, more or less in all parts of its walls. But it was very marked, more marked than anywhere else, in the fleshy columns, and this, perhaps, was one reason of the marked irregularity. They performed their office very imperfectly, and by sympathy disordered the heart.

This kind of degeneration is, again, of rather rare occurrence. It may occur at any age. It may occur upon hypertrophied and dilated hearts, and it may occur upon hearts that are perfectly normal in size, and up to the time of its occurrence perform their function properly.

I do not know what causes Quain's degeneration. It is something connected with the nutrition, the nerves of the organ. Persons who have it do not become œdematous, they do not become always very pale. This lady that I refer to was pale and icteric, but I did not refer the icteric element to the heart, unless, perhaps, the liver was deranged by imperfect circulation of blood through it.

As to the treatment of this particular form, if you make it out, there is but little. A rather extraordinary recommendation has been given, however, by one physician of some authority, and that is to advise the patient to run up-stairs. I think it would kill more than it would cure. The idea is to make the heart active, and to cause it to be nourished by healthy muscular tissue in this way. But my own opinion is that this cannot be done. The treatment that I should adopt, and have recommended in cases where this disease was suspected, is of an entirely different character. It is aimed at causing the absorption of the fatty matter, and reduction of the supply of fat in the body; for example, the patients, as in two or three other forms of disease that depend upon fatty disorders that I shall describe to you further on, are directed to take moderate exercise, to walk upon the level ground, or upon the floor of the house, and take, according to their strength, the exercise they can endure. The second element of the treatment is to exclude from the diet all fatty food, including butter, milk, cream, fat of meats, and in a manner all fatty food. I sometimes allow them enough of milk to flavor their tea or coffee, but nothing beyond that. Only lean meats, and the vegetable foods that contain the least oily matter. Well, the grains; perhaps you ask whether they contain oily matter; yes, they contain oily matter. Wheat contains about three per cent. Indian meal contains about ten per cent—that is the reason it fattens the hogs. Then I should prescribe food made from wheaten flour rather than from Indian meal. And next, and perhaps the most important element in treatment, is to administer the bicarbonate of soda, and to administer it in as large amount as will be borne, and that fact to be judged of by a daily examination of the urine. In giving alkalies you are to bear in mind that some of them have the power of producing calculi in the kidneys or bladder, but there is no danger of their producing anything of this sort as long as the urine is in the slightest degree acid. Furnish your patient, then, with slips of litmus paper, to be kept in a convenient place, as in a box about two and a half inches long and an inch and a half wide, and let the patient take every morning, or every evening (better in the evening), one of the papers by one of its ends and dip the other in the water, and when it comes out, notice whether it is red or purple. If it is purple the patient is taking too much soda; if it is in a very slight degree red, that is all right; if it is very red, he is not taking enough; give more. And in this way, from day to day, by applying the test, you may give what will be safely borne, and what will produce most effect upon fatty degeneration of any kind in the body of the patient. I do not know from experience that it will do any good in this kind of fatty degeneration; but it does so much good in other kinds of fatty degeneration that I deem it important that it be tried. Of course, the most nourishing food that the patient can take is to be preferred; beef and mutton, all lean, and not excluding chicken and fowls of various sorts. Any good wholesome meat food is better than vegetable food in this case; at the same time some vegetable food may be taken.

Before I go on with the matter of rupture of the heart it is probably best that I give you a few statistics from different authors referring to the frequency of disease in the several valves of the heart. Here, for example, in 300 cases of endocarditis in the adult, 297 were on the left side, 32 on the right side; 268 were confined to the left side, 3 confined to the right side; 29 in both. (Taken from statistics.) In the fœtus and newborn child, the evidence of endocarditis was in 192 cases in the right side, and in 15 in the left. Observe the very marked disproportion between the adult record and the infant record; 192 in the right side in the infant or newborn, 15 only in the left; in the adult 297 out of 300 cases on the left side; but 29 of these were on both sides at the same time. In regard to the particular valve that is most liable to disease, in the 300 cases 255 were in the mitral; a portion of these were in the aortic also, as you will see further on; 129 aortic; in the tricuspid valve, 29; in the pulmonary valve, 3; or one in a hundred. Of these 300, 159 were of the mitral only; 40 were of the aortic valve only; 3 of the tricuspid valve only; and none of the pulmonary valve only.

Here I have a record obtained from Bouillaud; and observe it is in hospital practice, for children in Paris are sent to a hospital separate from the general hospital. He found endocarditis occur in 55 per cent of cases of articular rheumatism. Budd, an English physician, finds it in 48. Fuller, in 23, Wunderlich and Lebert, in 23 per cent in a hospital for adults.

In regard to embolism, of which I have said something to you, in endocarditis or after it, different parts of the body are the seat of embolism, or parts that obstruct the flowing mass. In 84 cases that were observed in the Berlin anatomical museum, 57 were of the kidneys, but not of the kidneys alone; the kidneys were the seat of embolism in 57 out of the 84 cases. But these emboli are not always single; the spleen was the seat of the same accident in 39 of the cases; the brain in 15 cases; the liver and alimentary canal in five cases only out of 84; and the vessels of the skin in 14. Of course, in several of these the emboli were multiple; that is, different organs were affected at the same time.

LECTURE VIII.

RUPTURE OF THE HEART.

NOW, with reference to rupture of the heart. I do not believe the heart is ever ruptured when it is in a sound condition. You have just seen some illustrations of rupture of the heart, but bear in mind that some previous disease may have existed; it may be there was what they call myocarditis, a form of disease that I have not yet described to you. That produces, as you will see when I shall describe it, a weakening of the muscles here, and makes a place of least resistance, and a place that is liable to give way under the power of other portions of the heart. The specimens that you are examining show fatty degeneration, or rather adipose degeneration, a weakening of the muscular tissue, for fatty tissue has little consistency, gives but little resistance to the power that is acting on it. Now, if you had the means of determining in what particular cases this adipose degeneration of the heart has occurred, you might anticipate such an occurrence, or rather you might conjecture that it might occur. And that is about as far as you can go in the diagnosis. I told you when we were last together that we have no means of diagnosticating fatty degeneration of the heart. You observe in these plates that the organ is not enlarged. There is no valvular sound to indicate the presence of any new structure; therefore, auscultation amounts to nothing. The heart beats naturally and easily until the rupture comes. It may come during some period of considerable exertion, and the first you know of it is sudden death— and then you do not know what is the cause of death until you have made a post-mortem examination. So that speculation in this direction is really a loss of time. You can, in the language of Goldsmith, be sure, after she is dead, that her last disorder was mortal, and that is about all you can say. The accident may surprise you, then, at almost any time. Still, it is not a very frequent occurrence. When it occurs from a weakening of the walls of the heart, of the muscular tissue of the heart, from previous disease, other than fatty degeneration, it will be likely to take you by surprise equally, because there is no murmur, there is no friction sound, there is nothing that is unnatural in the action of the heart, unless perhaps, occasionally a little feebleness which you can account for in a dozen other ways.

When it follows myocarditis you have no means of anticipating it. It is, then, practically a post-mortem disease and you can prescribe no treatment.

Dr. Ferguson * reported to the N. Y. Pathological Society a rupture of the heart in a man fifty years of age. His health was good till a few hours before his death, when he suffered from dyspnœa. He died in the water closet, a linear rupture in the anterior wall of the left ventricle began half an inch above the apex and extended upward three quarters of an inch. There was an increase of the fibrous tissue and pus in the muscular wall of the ventricle.

Also another instance of rupture during a convulsion in tetanus. This was in a man forty-four years of age. He had a laceration of the left hand, and three weeks after tetanus began on the anterior of the heart an inch to the right of the apex there was a circular opening of one eighth of an inch in the right ventricle, the wall of which at this point was a line in thickness.

---

* The *Medical Record*, Nov. 24, 1883.

Dr. Da Costa * in his Toner Lectures relates a most extraordinary breakage. A man of forty years had paroxysms of severe cardiac pain happening at irregular intervals. They were attended by venous pulsation. This pulsation was noticed while he was suffering pain, but was marked only while this lasted, or just before or after. At this autopsy the heart was found to be enormously hypertrophied; one of the papillary muscles of the tricuspid valve was torn from its attachments both to the walls of the heart and to the tendinous cord, and a piece was found floating in the pulmonary artery. The right auricle and ventricle formed a continuous pouch that was filled with coagulum.

Dr. Coupland † reported a case of spontaneous rupture in a man aged sixty, from sudden fright. He says that perhaps the fullest information on this subject is given by Dr. Quain in his Lumleian Lectures, *Lancet*, 1872; 1. p. 460, where he states that in 77 out of 100 cases there was fatty degeneration of the heart walls and that 63 were over sixty years of age. He also points out what all writers have noticed, that the exciting cause of the rupture was some sudden mental excitement or physical exertion. This patient, till the accident occurred, had had no symptoms pointing to heart disease, yet the coronary arteries were advanced in calcareous degeneration, and there was considerable adiposity of the heart wall as well as true fatty degeneration. This case resembled the majority in the fact that the lesion was on the left side. In most cases the lesion is in the anterior wall, either in the region of the apex or close to the septum. In this case it is on the posterior wall, at some distance from the septum and about midway between the apex and the base. According to Dr. Quain 71 deaths in 100 were sudden, occurring in one or two

---

* Lecture III.   † *Medical News*, Dec. 30, 1882.

minutes, one patient, however, lived eight days; 1 five days; 1 nine days, etc. This patient lived forty-eight, and the inspection disclosed two clots in the pericardium of different ages, and between the two effusions he rallied considerably.

Dr. Van Santvoord* reported to the N. Y. Pathological Society a case in which death seemed to have been hastened by entrance of air into the pleural cavity in the operation of tapping. The heart was markedly hypertrophied and dilated. The pericardium was thickened and seemed to be undergoing atheromatous change. The valves were healthy. The microscopic examination of the fibres of muscle showed no degeneration on the left side. One exhibited a marked tendency to longitudinal splitting of the muscular fibres. On the right side the muscular fibres were in a state of moderate fatty degeneration. Why there was heart failure was not shown by the examination.

Dr. Displatz † relates a case in which a sharp fragment of bone lodged in the œsophagus about half way down. An abundant hemorrhage occurred on the eighth day which ceased spontaneously. Thirty-six hours later it was renewed and was followed by death. There was ulceration of the œsophagus and perforation of the aorta about an inch below the junction of the transverse and descending portions.

Dr. Wm. Axford ‡ reports two cases, one in a boy aged 16, the other in a boy of 15 years. The first "while carrying a heavy weight made a sudden and great muscular effort. Soon he complained of considerable pain in the centre of the sternum, also of uncomfortable feeling in the left arm; expectorated some blood; short dry cough." A

---

* *Medical Record*, March, 17, 1883.
† Ibid., June 2, 1883.
‡ Ibid., March 24, 1883.

time after his pulse was 100, temperature 97½°, respiration 26, irregular. There was an aortic regurgitant murmur. Since this accident the boy has not been able to do any hard work, but earns his living by bookkeeping.

In this second case, a boy in good health made a sudden and violent effort to control horses he was driving, and immediately felt pain in the region of the heart and said he could not get his breath for several days. Syncope and a sense of impending suffocation followed mental excitement or muscular effort often, and up to the time of the report he could not make any physical exertion. All who saw him found mitral regurgitation.

Dr. Axford adds that Corvisart was the first to point out the fact that this injury was possible and that it could by caused by muscular exertion. Dr. Peacock, he says, in 1865, had collected seventeen cases, four of which came under his own observation. The valves injured were: the aortic ten; the mitral four; the tricuspid three.

Dr. Brenner* reports the following: A woman fifty-five years old, had repeated hemorrhages from the stomach and died in consequence. She had had pleurisy of the left side six weeks previously. The left lung was retracted and closely adherent to the diaphragm. The two surfaces of the pericardium along the left border of the heart were closely adherent and the dilated stomach along the lesser curvature to the diaphragm. About two inches from the cardiac orifice on the lesser curvature was a round ulcer almost as large as the palm of the hand, at the edges of which were seen gaping coronary vessels. At the bottom of the ulcer was a sinus communicating directly with the cavity of the left ventricle through an opening in the endocardium the size of a pea. There were no serious cardiac disturbances during life. " The

---

* The *Medical Record*, Oct. 7, 1883.

hemorrhage was from the coronary arteries of the stomach and not from the heart."

In two cases of perforation of the left ventricle of the heart by gastric ulcer Oser,* at autopsy, discovered a round ulcer of the stomach which had penetrated the left ventricle of the heart. The communication between the organs was through a long narrow canal. No air was found in the heart or vessels. The woman was seventy-one years old. The perforation was indicated three days before death by the vomiting of bright arterial blood and by tarry stools.

Brenner had just published a similar case occurring in a woman 55 years old. She had had for years attacks of cardiac pain, occasionally attended by vomiting. A few days before death she vomited blood, had black tarry stools and great cardiac distress. A circular ulcer was found in the lesser curvature of the stomach, which communicated with an opening in the wall of the left ventricle.

Dr. Block† of Dantzig has been trying to show that the heart when wounded can be sutured and life saved by experiments on dogs and rabbits. Four experiments with rabbits showed that both thoracic and pericardial cavities can be opened for a short time with impunity. An opening of the right and left ventricle as well as an entire compression of the heart, for the application of suture, can also be supported by animals for a few minutes, and he presented a dog in good health in which there had been a wound of the cardiac muscle with opening and suture of the three thoracic cavities.

In order to prevent the escape of blood in the application of the sutures the heart is to be seized at the apex and drawn forward until pulsation and respiration cease

---

* *Am. Jour. of Med. Sci.*, April, 1882.
† Ibid., Jan., 1883.

(the animal not being necessarily killed by the procedure) or the traction on the heart can be made only sufficiently strong to arrest the escape of blood from the wound. The wound can then be ligatured or sutured.

LECTURE IX.

FIBROUS DEGENERATION OF THE HEART.

Now, then, I have something to say to you about the heart clot. No, not yet; fibrous degeneration of the heart comes before that, because it is of the heart structure itself. Dr. Quain, who gave us our first ideas of fatty degeneration of the muscular fibres of the heart, a few years ago, described what he regarded as hypertrophy of the heart arising from increase of its connective tissue. He called it fibrous hypertrophy. I have been, from the time he made this statement, a little skeptical in regard to the correctness of his opinion. But during the present winter I have examined by the microscope some specimens that are unquestionably of the character that he described, and a little further on, when you have fewer specimens to handle, I will show you some specimens under the microscope that will illustrate this form of disease. In specimens in my possession newly formed fibrous tissue or connective tissue is seen by the microscope in great abundance. In parts it is laid in between the muscular fibres of the organ. This new material, thus deposited, may have the width of one muscular fibre, or that of ten or twelve and more. In other parts it cuts across these fibres, causing their absorption so far as the new deposit extends. Then again it is laid in intermittently, leaving a short fragment of a muscular fibre isolated, and then another and another, all lying in the same line, contiguous fibres all broken up in the same way over a space which may fill two fields of the microscope. Thus, in the field, which seems to be four or five inches in diameter, fragments of muscle, appearing half an inch long, broken by a layer of connective tissue and disappearing, reappear at variable distances, presenting another semi-fragment, and so the field is checkered, the connective tissue showing its peculiar cells, and the muscular fragments their double striation. Thus in certain parts of the heart the muscular fibres are broken up and their continuous action wholly lost. It seems that the connective tissue can take on hypertrophy by itself alone. The quantity of it that may be found upon the spleen, sometimes called a cicatrix, is enormous, and it is perfectly well formed connective tissue, with the fibres and with the nuclei, and it may be half an inch in thickness. The connective tissue in the liver is multiplied very much in certain forms of disease, as is also the connective tissue in other parts of the body occasionally. But in the heart it seems to have had but a recent discovery. All the hypertrophies that have been noticed have been deemed to be hypertrophy of the muscular tissue only. As to the mode in which this particular form of disease presents itself under the microscope, when you make a section of an enlarged heart, you follow the muscular tissue a certain distance, and then it breaks up, and in the place of it you find connective tissue, and the connective tissue may be wide enough to reach the whole breadth of the field of the microscope, and no muscular tissue in it. But you see muscular fibres ending, one here and another there, then gone, and then by and by they will be observed again— that is, at a little distance from where they disappeared. It seems then to destroy the muscular tissue, when the hypertrophy of the connective tissue is marked. But frequently it seems to occur without being really a disorder. There is always between the layers of muscular fibre of the heart some connective tissue. The bundles are separated by something more than a myolemma, sarcolemma, or whatever name you may give it; some real connective tissue thrown in between the sarcolemma of the different bundles of fibres. You see that in passing a· microscope over a thin section of the heart in almost any instance. Well, this increases in quantity, of course increases in thickness, and by just so much as it increases beyond a certain point it seems to diminish the muscular fibres by compression; that is, causes their absorption and removal. When, then, this form of disease presents itself in any decided manner, you can easily comprehend that you will have an enlarged heart with diminished power; this deposit is for the most part in lieu of, or a substitution for, the muscular fibres, and as it has no contractile force, upon itself, it will only diminish the strength of the heart. I do not know that we are advanced far enough in our knowledge of disease to distinguish between fibrous hypertrophy and muscular hypertrophy; it requires some further observation; but it is important that you know that there is such a thing as fibrous degeneration of the heart with increase in size. It is a matter that is open to study yet. The study has not been pursued far enough to enable us to lay down rules and principles in regard to it.

LECTURE X.

HEART CLOTS.

Now, then, we come to heart clot. I give a little time to the matter in the lectures, as I used to receive from the graduates of the college long letters in regard to it. I remember one, four sheets of foolscap, written on each side of the page, describing a something that was terrible. A clot had formed in the heart, and extended out in the vessels about as far as the examiners could pursue them, and it was wonderful that a man could live with such a thing in his heart; well, he did not. All that formed after he died.

A distinguished physician of Philadelphia wrote a book on the frequent occurrence of ante-mortem heart clot. He thought they were of very frequent occurrence. I do not believe that. That clots can form in the heart previous to death, and may cause death, I fully believe, but the occurrence is rare. I will make you understand that, I think, before we get through. Now, to distinguish between the clot that is formed before death and one that is formed after, there are two or three rules that will be diagnostic. You come to a heart, the left ventricle of which is filled with coagulated blood, and the blood runs on out through the aorta, and you can follow it to some of the ramifications of the large vessels; or it forms in the right side, and extends through the pulmonary artery into even minute branches of the pulmonary vessels, and it does really look like a very formidable thing, and a person who does not understand it would assume that it was as bad as violent death. Now, then, if a clot form in the heart before the heart ceases to beat, it can never be attached to all the periphery of the heart. If the heart has ever beaten upon the clot that has formed in it, the contraction that must follow, if life continues, will separate the clot from all the external walls, from everything except the septum ventriculorum; and when clots are found in the heart that have been formed during life, they are always attached to the septum, usually to the lower part of the septum, where are the greatest number of pectinean muscles. The very act of contraction must separate it. If you will think of it for a moment, you will see it cannot be otherwise.

These clots are of two colors, buff and black. The buff is uppermost with reference to the position of the body after death; the dark colored is the dependent portion. Now what is that buffy coat? Exactly what takes place in the blood when withdrawn into a bowl and allowed to stand. The upper layer has been formed by the sinking of the blood corpuscles into the deeper layer, and that takes time. There is not time between two heart beats for anything of that kind to take place, and the occurrence of a clot with a buffy portion and a dark portion is conclusive evidence that the coagulation took place after death. The heart was full; it died expanded, and the coagulation took place in a manner that would require several minutes to form it, perhaps an hour or two. The blood does not coagulate quickly within the body. It has been estimated that it requires about six hours for the blood to become firmly coagulated after death. Sometimes it takes a much longer time than that, but a quick coagulation does not take place. As the blood-corpuscles are a little heavier under most circumstances than the fluid portion of the blood, their specific gravity being a little greater, they have time to sink to the deeper portions of the pool of blood, and allow the surface to coagulate with a buffy coat, a yellowish, semi-transparent substance. That, then, is conclusive evidence against the chances of ante-mortem heart clot.

It has been assumed frequently enough that the indentation of the clot at the point where the valves are placed is evidence of an impression made upon it by the action of the heart during life. But this is altogether another thing. When death takes place and the heart ceases to send a current of blood through the aorta, the upper portion of the valve of the aorta falls downward of its own weight into the blood; the lateral ones fall a little but not so much. The anterior is the point where quite an indentation in the coagulum has been noticed; it falls into the fluid blood, and when the coagulation takes place, it will take place about that valve that hangs in it, and of course it will leave an indentation in it. So far from being an evidence of ante-mortem clot, that particular thing is evidence that it formed after death. Almost all the clots that are found after death are formed after death in the heart and great vessels.

If the clot has formed before death, and the heart has beaten upon it, it has become compressed and will be found attached to the lower part of the septum ventriculorum and not to the walls of the heart, unless it be the very apex. Clots are formed in the heart. Here is a figure of one, and you observe it is rather crab-like in form; it has several legs.

Cardiac Thrombosis in Acute Disease.—Dr. Goodridge * bases his opinion on three cases in which death occurred with greater or less degree of suddenness and with attendant symptoms of dyspnœa, thoracic oppression, rapidly diminishing heart power, and a sense of impending dissolution. In these cases there were found firm decolorized

---

* *N. Y. Med. Journal*, Oct. 20, 1883.

coagula, occupying in one case the left ventricle and the beginning of the aorta; in another, exténding from the right ventricle into the pulmonary artery, and in the last blocking up the right auricle and the superior vena cava. In the last case the fibrous coagulum was distinctly laminated, and the symptoms pointed to a progressively increasing interference with the function of the right heart.

At a meeting of the section in Practice of the Academy of Medicine, Dr. J. Lewis Smith[*] defended the opinion always taught from this chair since I have occupied it, that the white clots and those that have an upper layer of white and a deeper one of very dark color are always and necessarily of post-mortem origin. He brought into his argument a very important case. He had in the hospital a man whose symptoms led him to suppose that the right heart was overloaded with blood. Feeling that benefit might follow the withdrawal of some of this blood he introduced a hypodermic syringe and removed a certain quantity. This blood was of dark color and perfectly fluid. The heart had ceased beating before the syringe was introduced, and cardiac action did not follow. Dr. Janeway then injected a solution of carbonate of ammonia into the same cavity, but it also failed to excite the heart to action. At the autopsy there was found a firm white clot entangled in the chordæ tendiniæ and columnæ carneæ and not any dark fluid. Immediately after the heart ceased to beat, the right cavity contained fluid dark blood. At the autopsy there was no fluid dark blood in the same cavity, but a firm white clot entangled, as described.

---

[*] The *Medical Record*, Jan. 13, 1883.

LECTURE XI.

VALVULAR DISEASE.

IT is necessary to refer to other causes of valvular diseases. The most common is the deposit of that substance that looks like mucine, that I have described to you. Its organization and contraction, producing what I have already denominated the shortening, thickening, and stiffening of the valve. .These same results may be obtained when other agents are at work. One of the things that operates most actively after inflammation is atheroma. Atheroma originally meant an abscess, but it has come to mean, in these later years, a particular deposit that is confined almost entirely to the heart and arteries, much more common in the arteries than in the heart itself, but in the valves not infrequently met with. This atheroma you can form an idea of by the specimens which I can exhibit to you. When you first find atheroma in the dead body it is almost always of a yellow color, or a faint yellow color. Here it is white. It has been bleached by alcohol, in which the specimen I show you has been kept. If you handle the specimen you will find that at different parts it is of uneven thickness. There are little warty, flattened eminences upon the whole circumference of it, and at the edge you can get an idea of its thickness. Here is another artery that is, to a certain extent, warty, exhibiting the same thing, lacking the yellow color which belongs to it when it is first seen. You observe it runs up the artery a considerable distance, that is the aorta. You observe, too, that it has narrowed the openings of the three arteries that are given off from the aorta. Here

VALVULAR DISEASE.

is an interesting specimen of an atheromatous deposit in a valve and not in the artery, or if any, very little. The valve has undergone one of the changes that I will describe to you directly, and it has ulcerated a hole through the valve. Not only that, but it seems to have been deposited upon the substance of the heart, probably not in the substance, and an opening has been made at the same time from one ventricle to the other. A broom straw has been passed through the whole, from one ventricle to the other, and you will observe what an ulcerated and damaged condition this valve is in. Here are a number of other specimens which you will examine.

Now, then, the first question that comes is, what is this— this atheromatous matter? Well, it is an effusion on the outside of the inner lining of the artery; the intima, as it is the fashion to call it now. It lies outside of the lining membrane of the artery. It is to a certain extent infiltrated into the fibrous tissue of the artery, but it is much more commonly found just between the lining membrane of the artery and the fibrinous layer, called media in these later days. Between the intima and the media, then, is to be found this deposit. And it is constituted sometimes of a material very much resembling that which I have described to you as thickening the valves in endocarditis, a sort of mucous matter, and that is thick set with cells that do not grow, do not develop themselves according to the rules of cell growth in other parts of the body, but become attached to each other on almost any point where they may touch, and thus form a kind of tissue. Then that is pretty fully infiltrated with little globules of oil, and occasionally a few fibres are formed by the coalescence of these cells, but they possess very little power of permanent life. And the result of this is a pretty early change in their constitution. The cells disintegrate, and oily globules seem to take their place.

The fibres gradually disappear, being broken down into granules, and soon the means of organization have passed. This takes place very soon, that is in a few months, or a year or two at most. The result of the breaking down of the structure is a softening, and you can sometimes find these little deposits of soft material, looking exactly like pus, except a little yellow. Put them under your microscope and you will find that there are no pus globules in them. The result of this is, sometimes, to leave an ulcer on the artery, or, as in the specimen you are examining, in the valves, and more frequently at the base of the valves than at any other part of them. A weakening of the structure is therefore produced, without softening. This atheromatous matter has no strength and it has no elasticity. It cannot replace in function, as it replaces in position, the fibrous tissue of the artery. The wall grows weak by the deposit of this matter in it, therefore, and you may start, with regard to aneurism, with this assertion, that aneurisms in the human body do not occur except by violence or by the weakening of the coats of the artery or through an atheromatous deposit.

An aneurism in any of the large trunks of the circulatory system will not occur without atheroma; the artery must be weakened before it will yield, and weakening is always effcted by atheroma of the artery. You may search for twenty years among aneurisms, and you will not find a single one that has occurred in a sound artery. Even those minute ones that have been lately described in the brain circulation have for their base atheromatous weakening of the arteries of the brain. Well, this is a pretty grave matter, then, for it is the parent of aneurism; it also produces the weakening that precedes what is called dissecting aneurism. The artery may break partly in two, the inner layers may be broken, and the blood may find its way between the layers of the artery for a very considerable distance. This, however, I believe only occurs in the large arteries, the thoracic and the abdominal. I do not know but it occurs in the iliac sometimes. That, also, I say, has its origin in the weakening of the arterial coats by this atheromatous deposit. This is one of the courses that an atheroma pursues when it is once deposited in the vessels. Another is interesting perhaps, not to say important, in its bearing. The atheromatous matter is broken down very much in the same way as I have described to you, but more gradually, and the fat which seems to be taken up by the circulatory vessels, and deposited in the place of the atheromatous matter is bony structure, bony scales. Such a specimen as that I will try to show you the next time we come together, in which the whole artery has been converted into scales, a multiplication of scales, some of which have a real bony structure—have the lacunæ and the canaliculi; and some of which are mere calcareous accretions, without anything which distinguishes bone. These render the artery inelastic, and so interfere with the circulation. As I have already told you, the heart sends the blood into the artery and dilates it, and the artery, by its natural contractile power, propagates or continues the action produced by the heart, carrying the blood forward in the circulation. But in an artery that has been deformed in this way the elasticity is gone. Hence, in such cases it is not uncommon that hypertrophy occurs. There is another evil that may come of these bony scales; they are sharp at the edge; they are thin. I say sharp; they have almost a cutting edge; and the gradual action of the artery, or so much as is left of it, may cut through the inner lining and let one of the sharp edges stand out like the edge of a nail, or the end of a nail rather, and on this may form the accretions of fibrine, the coagulations of fibrine, or vegetations, and these, in their turn, may wash off and make emboli. In a few instances, when the process is softening, and it has occurred between the valves, the soft matter may remain between the valves some time. I had a specimen—I do not know but I may have it now, but I have not seen it for two or three years—in which the mitral valve was the seat of atheromatous deposit, which made a material looking like pus. It was found after death. It at first merely separated the two folds of the mitral valve, and formed a kind of abscess. The quantity of softened material it contained was perhaps half a teaspoonful; it had not, as in the specimen you are examining, destroyed the lining membrane, or at least the folds of the valve.

There is one other point to be referred to in this connection. You hear a good deal, or something, at any rate, of bony deposits in the heart, and particularly in the valves. They almost always result from a changed atheroma. The atheromatous matter is deposited in the valves, and there is, perhaps, some thickening, too, from inflammation, produced by the irritation of this substance, and at length absorption of the material takes place and calcareous matter is deposited instead. I shall be able to show you, probably, two or three hearts, perhaps, some in which you can feel the bony roughness in the valves. Occasionally this bony deposit occurs in the heart itself.

I was consulted in regard to a physician who was pretty well advanced in life. It was supposed that he had heart disease. His heart was beating irregularly, sometimes palpitating, sometimes beating very feebly, and he was subject to fainting spells. On examining his heart I found no enlargement. I found nothing to indicate valvular disease, and I said. from the manner in which it behaved, there must be some disease of it, but I cannot tell what. In a month or two this gentleman died, and a post-mortem examination was made, and a mass of calcareous matter was found in the body of the heart, beginning at the base, and extending downward an inch or more, and, to a certain extent, thickening the particular part of the base of the heart in which it was deposited. It had crippled the heart in its muscular contractions, and the heart resented it as well as it could, but it could not tell the story to the outsider. It was plain that the heart was not working properly, but why, nobody could guess, because there are other forms of disease that behave in the same way, particularly, fatty degeneration, Quain's degeneration of the heart. You can frequently see that there is something wrong in the heart, but you cannot tell what it is. This atheromatous matter is not usually deposited in the muscular fibres of the heart, and whether it was the result of a previous deposit of this material in the heart, or whether it was a bony deposit or a calcareous deposit in connection with atheroma, I cannot tell. The importance of atheroma, then, will impress itself upon your minds mainly with reference to the deposit in the valves of the heart and the formation of calcareous or bony concretions in the valves, or, perhaps, in the substance of the organ. There is no way of distinguishing, before death these several forms of diseased valves. You cannot tell, as you hear a certain murmur, whether it is produced by inflammation of the pericardium or by an atheromatous deposit, or atheromatous deposits that have passed into calcareous or osseous formation. To determine this you will have to wait until after death, and as almost all, and probably all, the vegetations of the valves are incurable, to distinguish the varieties is of no practical importance.

 A remark or two more in regard to endocarditis before we pass to diseases of the valves to consider them in detail. Endocarditis, like pericarditis, is very apt to show itself to a certain degree upon the heart itself. An œdematous effusion may take place in the fibres of the heart exactly as you see an œdematous effusion occurring in the leg when the knee is the seat of inflammation, or in the arm when the muscles of the shoulder are involved in inflammatory disease, as rheumatism. For example, in February of last year I woke one morning with a pain and stiffness in the left shoulder. I found that I could not get my coat on without assistance. So I got assistance to put on my coat and went about my business. The arm, however, grew worse and worse, and I recognized after a few hours that I had rheumatism in the deltoid muscle, and directly my arm and hand began to swell and my hand looked very fat. As soon as I saw it swelling I took off my rings for fear they would be fastened upon the finger and could not be gotten off. The swelling continued until, after a pretty free use of the bicarbonate of soda, the inflammation was subdued. Then the swelling all went away. So in the heart, the parts neighboring to those that are the seat of inflammation may become the seat of œdematous effusion, and thus weaken the heart—the same thing that I stated to you when speaking of pericarditis. Pericarditis and endocarditis commonly occur together, and of course there will be on this account all the more of this œdema of the heart.

Dr. Livingstone does not believe the little nodules on the edges of the mitral and tricuspid valves are of inflammatory origin, for out of 136 autopsies of children either stillborn or a few hours old up to three and a half years of age, he had never failed to find them *when he had looked for them*, and he says they are the more pronounced the younger the child. He does not admit that they result from the rupture of the intravalvular blood-vessels, causing hæmatomata from which the blood has been absorbed and only a semi-transparent substance remaining. But he is inclined to the idea of Richard Pott, that they are the remains of the fœtal valve, whatever that may be. He says: "They are composed of a collection of the normal elements of the valve and do not appear microscopically like the deposits of an endocarditis."

Dr. Livingstone raises the question whether the nodules on the mitral valve or the movement of the blood through the ductus arteriosus produced the systolic murmur heard at the apex, or whether it was produced at foramen ovale, citing de Gassicourt as authority for the idea that the current through the ductus arteriosus can cause a murmur.

Endocarditis is perhaps the most prolific cause of derangement of the valves, and it may occur as the result of acute disease commonly associated with rheumatism, occasionally found in connection with Bright's disease, sometimes occurring in the course of measles, and rather frequently in the dropsical sequelæ of scarlet fever, sometimes occurring in septicæmia. And still again, I told you it may occur independently of any of these causes as a spontaneous, I might say almost idiopathic disease. That it does occur frequently in this manner is rendered evident by the large number of persons that suffer from disease of the valves who have not had any sickness preceding the development of heart disease; who have not had rheumatism; who do not remember that they have had scarlet fever; who have not had Bright's disease, etc. I have told you of atheroma, a particular deposit, more frequently found on the inner, or rather under the inner, lining of the arteries encroaching upon the elastic tissues of the walls. I told you that it undergoes, whether in the valves or in the artery, one of two changes as a rule. It may soften and form a yellow collection looking like an abscess, and this may rupture the intima and escape into the blood, and do mischief or not according to the fineness of its division. Or it may harden and undergo calcareous degeneration. It does not seem possible that the elements of atheroma can be converted into bony matter. It would seem altogether likely, therefore, that it is substituted for these after they are softened and have been absorbed. I told you that this deposit in the valves causes a thickening, an inflammation, a bony deposit especially causing constant irritation. I told you that the valves are sometimes ruptured, and I pointed out to you a specimen in which the inferior or lower portion had given way, so as to render that part of the valve useless in retaining the blood in the artery. I told you of a case in which a man, while lifting a heavy beam, felt a sudden something give way, he could hardly tell what, but he said there was something gave way, and he fainted, and sat for a considerable time in the mill before he attempted to go home, and then reached home only with assistance. In a few days he revived again and went to his work. He lived twelve years after that with the heart largely hypertrophied. At the post-mortem the longest portion of the tricuspid valve was found to have been flapping backward and forward in the current of blood. As the blood ran down, the strand was in the ventricle, and when the heart contracted it would double back on the rest of the valve. A tendinous cord had given way when he had felt the sensation described. He had a murmur heard in systole; not in diastole.

The valve may give way in a considerable variety of cases. I recall one of a gardener who had loaded his cart too heavily. His horse could not draw it, and he put his shoulder to the rear of it to help the horse, and exerting about all his strength, he also felt something give way and was very much crippled after that. He came to Bellevue Hospital. There I had an opportunity of examining him before and after death. A portion of one of the folds of the aortic valve had given way in his case and I thought almost without previous disease. Atheroma is a very common precedent of the breaking of the valves. This man lived about two years after the accident occurred to him.

I once was with a gentleman who was making great haste to reach the top of a hill. He was within about fifty feet of it, when he started on a run up the steep acclivity. The result was he was brought into the same condition as the gardener, and from that time on, as long as I knew him, there was a regurgitant murmur at the aortic opening. He broke something at that time. Precisely what, whether the bottom of the valve gave way, or whether it split down, I cannot tell. One thing, however with regard to the splitting of a valve is worthy of notice. You remember there is on top of the valve a kind of cord, and immediately below that cord the valve is not as strong as at the edge. The common way for a valve to break is to give way at the cord, and it vibrates in the current of blood, and frequently makes a musical sound, a sort of Æolian harp sound; making a noise that is not altogether unlike that of a cord stretched across a little opening in a window. It is quite a musical note, and I don't know that there are any other circumstances in which that musical sound is produced. Then I should add, in this enumeration, that the tendinous cords are occasionally found welded together, two or three or more of them, and, of course, a great deal crippled in their action. In certain instances the two folds of the mitral valve and the three portions of the aortic valve grow together, in the latter case leaving a slit by which the blood can go out and by which it will return in regurgitation. How in the world that can occur I do not know, but it does occur. They are almost always thickened, so that they do not play freely before this adhesion or coalescence takes place. But moved, as they are, con-

VALVULAR DISEASE. 173

tinuously, by the current and reflux of blood, it is very difficult to understand how these three portions of the aortic valve can grow together, leaving only a little slit for the blood to pass through. It is easier to understand how adhesion may take place between the two curtains of the mitral valve, for by contraction they can be grown together, and contraction is affected by the material that is deposited between the two folds of the valve.

I will put in your hands, now, for example, a few specimens of valvular disease. Here is one, in which, as you open it you see that the wall is the seat of spots of atheroma, and some of these have passed into a calcareous condition. Even the valve itself is shortened and thickened, but not remarkably. In this one the aortic valve is very much diseased. If you put your finger on it you will find that it has hard scales on it. It is not very much contracted, except one fold of it. One fold is thickened and hardened, and on the inside of it is a little calcareous matter. This has been a pericarditis, and the outside of the heart is roughened by the adhesions that formerly existed on the pericardium. The mitral valve is in a state of very marked stenosis, that is, contracted, and you see here almost nothing of the tendinous cords, they are shortened so much. The fleshy columns are a little elongated, to make compensation for the shortening of the tendinous cords. This is a striking specimen of stenosis of the mitral valve, and in this connection it may be that we shall find an illustration of the rule that applies to such cases generally—that is, stenosis of the mitral valve induces hypertrophy of the right ventricle; stenosis of the left auriculo-ventricular valve induces hypertrophy of the right ventricle, and frequently of the right auricle, and also sometimes of the left auricle. This you see illustrated here. The right ventricle is thicker than natural, but not very much hypertrophied.

This stenosis, marked as it is, did not disturb the circulation as it sometimes does. Then in this specimen you have two lesions, the aortic valves thickened and a little calcareous deposit in them; and at the mitral valves, stenosis.

In this specimen the mitral valve is thickened and hard, but not disorganized, and it is large enough to allow the circulation to go on with ordinary freedom. The aortic valves are thickened, shortened, and hardened. In this specimen the aortic valves are a little but not much diseased. The chief lesion is probably in the mitral. You observe the mitral valve is thickened, not exactly disorganized. There is some hypertrophy of the heart. This is not a very striking specimen.

Here is a very good one, and opened in such a way that you can see the change that has taken place; it is in the aortic valves. You observe they are contracted in their length, contracted in height, and thickened, but not nearly as much as in some specimens I shall show you hereafter.

Now comes the question, what sort of disturbance will these lesions of the valves produce? Of course it will be first felt in the circulation. Let us take a mitral stenosis, such as you have in two or three of the specimens now going round, and its effects are perhaps as simple as the effects of any of these lesions. Indeed, they all follow in the same track after they have taken the first step. Remember, now, that the blood in the left auricle has just come from the lungs, and it is seeking its way into the left ventricle. It is obstructed at the valves; well, what will that do? The circulation by the right heart we are to assume is natural. The blood goes into the general system, and empties into the right side of the heart, and from that is thrown into the lungs, and it begins to come from the lungs into the left heart and is obstructed.

What then? The first thing will be a certain amount of engorgement of the lungs. They receive blood faster than they can discharge it; the consequence is, therefore, an engorgement of the lungs, greater or less, depending upon the extent of the stenosis. Next after that comes a strain upon the right heart, which is attempting to send blood all the time into the lungs, and associated with this may be what is called an emphasized condition of the second sound in the pulmonary artery. That is pretty full of blood, the right heart pumps more into it, the reaction is stronger, and makes the valves strike with more force than natural. Hence, there may be an increase in the second sound from the pulmonary artery, in an instance where the disease is at all considerable. Well, what more? Stop the blood from coming into the right heart and what will be flooded? Why, every important organ of the body. The blood cannot return from the veins with anything like its natural freedom. The liver will be engorged, the spleen will be engorged, the brain will contain more blood than it should, the face perhaps will be a little puffy, from a little effusion into the tissues from the veins. The veins cannot empty themselves freely. Little clusters of enlarged veins are found upon the surface of the body, chiefly upon the chest. The external thoracic veins frequently become enlarged. There is general hindrance of the blood in the general circulation, because there is obstruction of the blood in the lungs, and consequently obstruction in the right heart, to which the blood in the general system should naturally go. One consequence of this accident, then, is enlargement of the liver; not in all cases, not in the majority of cases, but now and then. And sometimes quite considerable hypertrophy of the heart, and enlargement of the spleen, from the same cause, namely, obstructed return of blood. The effect upon the stomach is often noticeable. There being more blood in the stomach mucous membrane, the digestive membrane, than can be used, the digestion becomes impaired. It is imperfect, and the appetite feels the effect of this. Then, too, the effect of the obstruction is sometimes so considerable as to hinder the contribution of the thoracic duct to the general nutrition of the body. The thoracic duct, you remember, receives the material that is digested in the intestine, and carries it to a vein through a duct, and that vein delivers it to the general circulation. Well, that is obstructed also; it cannot freely give its nutritious fluid to the blood, and it not infrequently happens that the patient becomes pale, in consequence of imperfect nourishment, though he may eat a considerable quantity of food; and you will observe, frequently—I am referring now to an advanced case—that the patient speaks of his feet being swelled, particularly at night, if he is on his feet during the day, and you observe yourself some puffiness about the face. You will particularly be able to notice what has been called the tear line; that is, a something looking like a silver thread or a thread of transparent glass lying upon the upper edge of the lower eyelid. It is a little œdema of the conjunctival covering, and as the eyelid presses considerably upon the eye it is forced up along the ridge, above the tear line. There may not only be œdema of different parts of the body, but the whole body may be swollen. The kidneys have suffered congestion, and have taken on a condition analogous to that of Bright's disease, generally the large white kindey, and they are unable to perform their function properly. The urea remains in the blood, or a part of it, at any rate, and produces its systemic effects. The urine is commonly scanty, sometimes even bloody, and there is not unfrequently effusion found in the chest, a double pleuritic effusion. I recently examined a gentleman from the country, or rather from a distant city, who had been ill for some months, and in the course of the examination found fluid effusion coming nearly up to the lower angle of the scapula on both sides. My thought was, he has Bright's disease. He had heart disease, for his heart was enlarged, but there was no valvular murmur at the time I examined him. On examining the water I found it contained a large quantity of albumen. It was one of the cases that interested the late Dr. Stephens so much when this matter was first investigated. There was a large amount of amorphous urates in his urine. On heating it up to about the point of boiling, it became cloudy with albumen, and heating it till it boiled it became almost thick. I have a scale for estimating the amount of albumen in the urine, ranging from one to ten. Ten represents a coagulation of albumen in the tube, so that nothing will run out; invert the tube and nothing is discharged; the whole of the urine is imprisoned in the coagulated albumen. When there is but a trace of albumen it is represented by one. In the specimen of the patient spoken of the amount of albumen was represented as six; or more than half of the urine was composed of albumen. I did not know that this gentlemen had Bright's disease when I found a double pleuritic effusion (although I felt pretty sure of it), for double pleuritic effusion does once in a while occur with Bright's disease.

This, then, is substantially the series of consequences that will result from obstruction, from stenosis of the mitral valve. And you can see that when the blood is returned through this valve, that is, when it has entered the ventricle, and a part of it is returned into the auricle, that it is going to do exactly the same thing when the condition becomes marked. It it all the same, save that in the one case it is stenosis, in the other it is insufficiency. The valve is incapable of serving as a gate, a watch, and a part of the blood that enters the ventricle goes back into the left auricle, and obstructs the blood that comes in from the lung, so that the result in the end is exactly the same—congestion of these various parts, sometimes cyanosis. This latter belongs also to mitral stenosis.

Then there is another point which I may as well refer to here. You sometimes see a pulsation in the veins of the neck, the jugular. There are two modes in which what appears a little like a pulsation is to be accounted for. When this obstructed circulation occurs the vein is seen to fill in expiration. In expiration the blood does not flow so freely in the lungs or into the chest, as in inspiration, and in expiration you may watch the vein and see it fill, but it fills from above. It is not a wave coming up from below. About the only conditions in which the wave will come up from below are when there is insufficiency of the tricuspid valve, that is, the valve of the right side of the heart, or when there is hypertrophy, a tumor like an aneurism resting upon the vena cava descendens. In either of these cases a regular wave may come up the vein, but I know of no other condition in which it does occur; but in the cases referred to you observe the filling of the vein is a different thing from an aneurismal tumor. It is not necessary to dwell longer upon regurgitation at the mitral valve, as the effects of it in the end are the same as those of stenosis.

LECTURE XII.

VALVULAR DISEASE—(*Continued*).

Now turn to the aortic valve, to the two conditions that have been named—obstructive disease and regurgitative, or direct and indirect, if you choose to so call it. When there is obstruction at the aortic valve the sound produced by it will be heard in the contraction of the heart in systole, and it will be loudest at the base of the heart, near the sternum. This may exist in a moderate degree for a long time, and no very grave consequences result. But if it is a growing disease, if it is a form of valvular disease that is irritating in itself, like a spicula of calcareous matter in the valve, action of the valve will excite an inflammation, and cause a deposit of material that can contract and still further deform the valve. But there are a great many persons who can carry both mitral regurgitation and aortic obstruction a very long term of years. Indeed, my colleague, the late Dr. Gilman, came to me with mitral disease at a certain time, and he said, "Am I going to die of it?" "Well, maybe," said I, "but I think you will live twenty years." The years rolled on, and I heard nothing more of it, until at length Dr. Gilman came to my house, and exclaimed: "Doctor, my lease is out! I want to renew it!" He lived about four years after this.

I remember a patient who carried a valvular lesion fifty years before a fatal sickness occurred; a young man, at first, all activity, but, of course, he could not go up-hill or up-stairs as well as other people, but his mind was active, and as an evidence of his industry, in the old time, when speculation did not fill a man's pocket in a day, he had made by speculating in oil a million of dollars, and built up the fortunes of two brothers in this period of fifty years.

I am in the habit, in this connection, of referring to a certain case because it is rather impressive. Many years ago I had occasion to listen over a certain gentleman's heart, and I found a mitral murmur. He appeared perfectly well, and I went on, making no sign of surprise, and afterwards thought the thing over. He is of phlegmatic temperament, and I do not think he will be very much excited; he will take the world pretty easy; I think he will take the world quite as easy without as with a knowledge of the disease of his heart. Now that gentleman is in active practice to-day; you see his name not infrequently in the medical journals, as having performed this or that operation, and he is a great deal happier than he would be if he knew that he had mitral regurgitation. There is no knowing, then, when the unfavorable issue will take place; but I can tell you this about it: when a patient comes to you complaining of a great deal of shortness of breath, when he has œdema of the feet, and possibly some swelling of the face, look to the kidneys. They probably have become involved by that time, and in a few weeks, in a week or two, he may be pretty gravely sick, having carried disease of the heart for twenty, thirty, forty, or possibly, fifty years, as in the case I have just referred to. And here it is perhaps proper to say that the last generation of doctors, and the present generation of people who are not doctors, look upon the statement that they have heart disease as equivalent to signing their death warrant. The profession have outgrown the idea that heart disease must be fatal in a few months; but the older physicians, physicians of past generations, not having auscultation to guide them, were not able to detect diseases of the heart until the kidneys, and perhaps liver, and possibly the spleen were involved. In other words, not until œdema came. Their post-mortem examinations told them what œdema meant. They were able to recognize then, by the general signs, what was the trouble, but the patients generally died pretty soon, in a few months after the œdema occurred. But they may have passed forty, fifty, or even sixty years before that, with the same disease of the heart, only it had not until the œdema appeared, come to be obstructive to the venous circulation.

But to go on with the obstruction at the aortic opening. If it is moderate there is compensation. The heart has the faculty of growing in size and in strength to meet an emergency, and we speak of compensating hypertrophy of the heart. If the blood does not flow easily through the aortic opening, the heart will gain additional strength by gaining additional fibres, and will overcome the obstacle, so that a man may carry a disease of that kind a great while without knowing that he has it. But when regurgitation occurs, and that is only in the advanced stages of obstructive disease, when the valves become shortened and thickened, so that they cannot fall together, a certain quantity of blood that has been sent into the aorta goes back again into the ventricle every time the heart dilates, and here is work to be done over and over again. The heart has to do its work twice over, substantially; at any rate, once and a half. The ventricle has just been filled from blood forced in from the auricle, and half filled from that which comes from the aorta. The effect of this is exactly that which I have been describing to you, and you will see how it must be so. When blood from the aorta falls back into the left ventricle it prevents blood from coming in from the left auricle. Only about half the contents of the ventricles can come in from the left auricle, and consequently the blood will set back upon the right heart, and engorge the lungs, and it will also produce very much the same general symptoms as those I have described to you. Well, then, these all come to about the same thing. Their consequences are substantially the same. So also in contraction of the mitral valve. A contraction of the aortic valve, you observe, is compensated for very generally, and that is not so grave a trouble. A stronger current of blood is sent over it, and of course into the general circulation more blood is sent in a given time, and consequently the general circulation does not feel it so much. These are the general effects relating to the advanced stage of cardiac disease.

The question has been often raised and discussed, which comes first, the kidney disease or the cardiac disease? My own answer is very positive: that while the cardiac disease may come after the kidney affection has begun, in the great majority of instances in which they are connected, the connection is as I have been describing to you. The kidneys become congested because the circulation is obstructed at the heart, and after they have suffered a certain length of time they will begin to give albuminous urine; they do not secrete the urea in full measure, and œdema then begins in different parts of the body. It is my strong conviction that in nine cases out of ten where the two are associated the cardiac disease came first, and the other case will be an instance in which kidney disease from some other cause has occurred, and the cardiac disease has followed it, possibly by the influence of congestion in a certain degree. But the order is almost always cardiac disease first, and renal disease afterward. I have noticed so often what I have already explained to you that I cannot have a doubt about it. I examine a man and find he has valvular lesion of one sort or another; I watch him for a year; he gets along very well; and perhaps ten, fifteen, or twenty years after I have examined him for the first time he begins to suffer; he begins to get œdema of the lungs, which is one of the effects I have not mentioned. We may have an œdematous condition of the lungs, as well as of the legs, body, and face.

With reference to the right side of the heart, we have had but little experience with disease of the valves on this side. For forty-nine cases of valvular disease of the left - heart there will hardly be one of valvular disease of the right heart. I have told you that after birth the left side of the heart is the seat of disease, and before birth the right. In the right heart there are lesions of the valves, though they are rarely met with. In a little boy about eight years old, when the college was in Crosby Street, I found a very peculiar sound in listening over the base of the heart. It was not a murmur, but it was a snap, and it was heard in both actions of the heart. It was a double snap, very much such as you get by pulling a piece of ribbon suddenly. He died, and at the post-mortem examination that sealing together of the pulmonary valves was found which I have already described to you. The edges of the three portions of the valve were sealed together, with the exception of an opening, which must have been less than half an inch in the middle, and one of the lips of that opening was a little longer than the other, and it was this which snapped when the blood came against it from the heart, and snapped again when the pressure from the artery was felt. That was the only lesion about it. I did not distinguish that from the aortic disease, for it is a difficult thing to do it, but some persons claim to be able to do so by getting the sounds of he pulmonary valves a little to the left of the sounds that come from the aorta. I dare not trust my ear for the diagnosis; and besides, there is another mode that is worth something, but not worth a great deal either, and that is, if the sound is at the base of the heart, to follow it to the artery, and particularly to the left and right clavicle. Listen to the sound from the heart there; if it comes up there pretty fully, it is in all probability in the aorta. I say, in all probability. I make that modification because the two vessels, the pulmonary artery and the aorta, are sealed together by loose connective tissue, and for a certain distance after they leave the heart they run together.

I have some curiosities here, some that are so rare that you may never see anything like them, and some that you will see plenty of. Here, for example, is a heart, the pericardium of which is attached to the heart, but between the pericardium and the heart is a layer of bone that makes a perfect hoop, or did make a perfect hoop about the heart. This bony material, you observe, is covered over by the pericardium on the outside, and attached to the heart on the inside. It flattens out to make quite a broad covering, then it grows narrow, and is broken, and you can feel the rough bone on each side—it then goes on up to the point of starting. How a heart hooped in this way could pump out blood is a matter of wonder. It was probably by a change in its usual motion. The contraction was probably from the point towards the base. There is also in this heart a considerable amount of thickening and hardening of the aortic valves, but not so much as in some specimens I shall show you. The amount of effusion that took place during the pericarditis, between the pericardium and the heart, was too large to undergo organization, and for some reason bony matter has taken its place. It was very likely broken down and absorbed in part, and calcareous matter deposited in its place.

Here is a heart that I shall have occasion to refer to, perhaps, at our next meeting. Observe, particularly, its flabbiness, and its color, which is yellow. This color is not so very deeply marked, however. It is a fatty heart. It is an example of Quain's degeneration. It is a feeble heart also; the walls are hardly of the usual thickness. But I will touch on this again further on. I have but little more to say regarding valvular diseases. I told you that the valves are occasionally torn out; the aortic valves are torn out at the bottom, at the attachment to the aorta. The cup is made no cup by having its bottom fall out, so to speak. That is almost always the consequence of atheromatous degeneration, weakening the valve at that point. There is another atheromatous change that is met with occasionally, and that is atheromatous deposit at the point where two of these valves are together attached to the aorta. You will observe as you examine these specimens that each pair of the valves, each of the two valves that are near to each other, have their attachment to the aorta at the same position; that is, one is in immediate apposition to the other. Now, it sometimes happens that the aorta at that point becomes atheromatous, and that the attachment is torn down. It is torn down so that it comes to be about a quarter of an inch below the attachment of the other valve, and this produces insufficiency. The valves will not, even supposing they are not diseased, meet in the middle of the aorta; one will fall in below the others, and consequently the valve is incomplete.

These, I think, with those enumerated before, will constitute the changes which you will be most likely to meet. Possibly not all the changes, but they are all that occur to me as occurring with any degree of frequency.

, There is a point which I omitted to refer to, but was reminded of by one of the students, and that is with regard to the pulse in regurgitation at the aortic opening. There is a peculiar pulse. It has not, as our Methodist friends say, *the gift of continuance*. It is a blow, short, sharp, I cannot say decisive, and then it seems as if the pulse is instantly cut off as soon as you feel it. There is a little duration to the swelling of the artery in the healthy condition, and still more in hypertrophy of the heart. Supposing there is no obstacle, but if there is regurgitation at the same time, the pulse has that short, quick, sudden beat aud sudden subsidence; the force that created the pulse seems to be withdrawn instantly after you feel the pulse, and then there is not infrequently, on account of the smaller wave of blood that is sent out, a little retardation; that is, the pulse does not come to the finger as soon after the heart-beat as in health. But this you can only appreciate by putting one hand on the præcordial region and the other on the pulse. The arteries in the neck, in instances of hypertrophy and regurgitation at the aortic opening, are apt to be enlarged and to have a wide beat. You observe I do not say *stronger* beat, but a wider beat than usual, and the pulse in them is apt to be feeble; and further, the arteries in time not infrequently grow tortuous—they wind about instead of going direct to the point of distribution.

From a number of examinations of hypertrophied hearts, with valvular lesions, Dr. Putjatin* concludes that in chronic affection the nerve ganglia are affected by an inflammatory process. In early and slight disease the changes are limited to hyperæmia and granular degeneration. In chronic and extended disease of the heart the connective tissue is produced and there is fatty and pigment degeneration of ganglion cells. In one case there was entire destruction of the ganglion cells and calcifica-

---

* The *Medical Record*, July 7, 1883.

tion of the tissues between them. Changes in the ganglia may also be produced by chronic constitutional disease. In this fact may be found the explanation of the function disturbances and even paralysis of the heart.

Another observer found that in hypertrophy with chronic interstitial nephritis with the growth of the muscular tissue the medullary sheath of the nerves is lost, and a process of nuclear proliferation commences. Should an acute disease supervene, then the change in the affected nerves assumes the character of an acute parenchymatous inflammation. In the nerve cells lying in the course of the fibres as well as those in the ganglia of the septum arteriosus the changes were confined to thickening of the capsule and increase of the nuclei. The protoplasm itself was not affected.

As bearing upon valvular disease of the heart both in reference to etiology, pathology, diagnosis, and treatment, I have collected several cases, some from those which have come under my own observation and others which have been reported in current medical literature, and shall also enumerate various experiments which have been made, and conditions which have been observed in valvular diseases of the heart.

Dr. C. S. M., seventy years of age, applied to me forty years ago for advice regarding a disorderly heart. I found that he had moderate enlargement and a murmur, I think mitral chiefly, because he has now that murmur feebly, and no murmur at the aortic opening. He was advised to eat moderately, not to walk up hill when he could ride, and when obliged to do this to take plenty of time, and go very slowly; as his profession would compel him to ascend stairs many times a day, to apply the rule of moderation with great rigor to this part of his duty. He has lived in obedience of these rules. He has taken no intoxicating liquors except when needed as medicine. He has had nearly all his professional life a large practice, but as soon as he had won the confidence of his patients sufficiently to do so he refused to go out after bedtime, and has long had a partner on whom that work has devolved. He has, until lately, appeared like a healthy man, and it is only in the last four months that he has not been able to follow his practice. A month and a half ago he came to my house suffering from shortness of breath, and having some œdema of the legs, and fluid in each pleural cavity, filling one half its capacity. He took diuretics. The kidney secretion soon amounted to seventy ounces a day, and he improved rapidly, so that he felt that he was all but restored to health, and ceased to take any medicine. He went from home, as he says, "on a summer vacation." While away the urine diminished in quantity to three or four ounces, was of very high color, and highly albuminous. With this came convulsions, and he had of these in a day and a half they say thirty, none very violent, but all with a short period of unconsciousness. With a digitalis poultice over the loins and other medicines the urinary secretion was restored to a healthy quantity and more, and he has had no more convulsions. After recovering some strength he returned to his home, where I saw him to-day (Aug. 23, 1883).

He has now no œdema at all. The fluid is wholly removed from the chest, but he has lost flesh since I saw him a month and a half ago, yet for most of the time he has taken about his usual rations, avoiding meat since the convulsions, but taking milk and eggs. The urine amounts to thirty ounces and over 'daily, and he says he feels well but is not strong. He walks the floor, but is not willing to go down stairs, fearing he cannot get back Albumen had not been found when I saw him first, and no casts; but I felt certain of Bright's disease because it so often follows heart disease even forty years after the discovery of the heart affection; and because there was double pleural serous effusion, without pain, or other evidences of pleurisy, and because with these there was œdema of the legs. I did not examine the urine, but others have; and since his first call on me albumen has been found, uniformly, more or less varying, and casts not uniformly, but often, mostly hyaline.

Now comes the peculiarity, to record which I have written up the case. When I saw the doctor in July I made out a double mitral murmur, which was not very distinct to-day, but his pulse at the wrist was *twenty-eight beats in the minute*, and no faint indication even of any more. But when the impulses of the heart were counted by putting the ear over it, there were twenty-eight pulsations that were distinct and strong, and twenty-eight more alternate ones which could just be heard and no more, and which sent no pulse to the wrist. Occasionally the feeble heart pulse would be strong and forcible, but for almost the whole of a minute it was the faint and the strong alternately.

Dr. Herman Tittenger[*] describes a peculiar murmur characterized by the distance at which it is audible (it can be distinctly heard through the bed covering) by its stability (it is but slightly influenced by the intensity of the heart's action) and by its character. Together with the murmur is perceived a widely propagated tremor. This phenomenon occurs only with rupture of the valves or tendinous cords. The primary murmur is caused by insufficiency of the valves, the secondary sound by the flapping of the loose valves and tendinous cords in the returning blood stream. A diastolic murmur indicates the rupture of the semi-lunar valves, while a systolic

---

[*] *Medical Record*, Aug. 18, 1883, p. 184, from *Centralb. für Klin. Med.*, June 9, 1883.

sound points to the (much less frequent) lesion of the mitral valve.

Aneurismal Dilatation of the Heart and Mitral Stenosis—Fibroid Induration.*—The heart is large; the anterior wall of the left ventricle is bound to the pericardium by a firm band of connective tissue. The left ventricle is enormously dilated, and the left ventricular wall in places is less than one eighth of an inch in thickness, and at the apex there is a portion of the wall where the pericardium and endocardium are separated only by the interposition of a thin layer of fat. Posteriorly the ventricular wall is much thicker, being uniformly nearly one inch. The papillary muscles are much hypertrophied. The right ventricle is moderately dilated, and the valves of that side are normal. The segments of the aortic valve are slightly thickened and retracted. The stenosis of the mitral valve only admits a cylinder of half an inch diameter. The endocardium is markedly thickened, and in all the ventricular wall there is increase of the fibrous tissue between the muscular cells (fibroid induration of the heart) especially marked where the ventricular wall is thinnest, except at the apex, where there is no muscular tissue. There is atrophy of the fibre cells, and much pigment arranged around the poles of their nuclei. The cells in places contain a small amount of fat, but generally the transverse striæ are well preserved.

Dr. Ferguson, who reported this case to the N. Y. Pathological Society, May 23, 1883, has examined the most of it. He finds parts hypertrophied, parts dilated, in parts fibroid substitution for muscular fibres, and here and there a little fatty degeneration of the muscular fibres. The patient was twenty-seven years of age; had had rheumatism. His chief suffering was from dyspnœa

---

* *Med. Rec.*, Sept. 8, 1883.

and palpitation, cough and expectoration, the latter scanty and sometimes streaked with blood. There was a double mitral murmur. Pulse 80 to 140, temp. 99° to 102°. He died unexpectedly. A *complete* diagnosis was impossible.

Dr. Ferguson exhibited at the meeting, June 27, 1883, a heart resembling this with aneurismal clots adherent near the apex in the left ventricle, wall there thin and wholly fibrous, muscular tissue gone.

Tricuspid Insufficiency.—M. François Franck,* by a specially contrived valvulotome, has produced tricuspid insufficiency and has noted the effects. 1. Systolic pulsation in the jugular veins and in the thoracic veins, dilatation and hepatic pulsation. 2. Systolic murmur in the heart, whose pitch is in inverse proportion to the extent of the sound; increased frequency of cardiac action; diminution in the arterial pressure. 3. Increased frequency of respiration; and 4. After a time the phenomena of general venous stasis?—ascites, anasarca, albuminuria, etc.

M. P. Dureoziez concludes that a tricuspid regurgitant murmur can be heard over the entire anterior surface of the heart, the points of maximum intensity being sometimes at the xiphoid cartilage, sometimes at the cardiac apex, and it may be heard at the latter spot only. It is not, however, transmitted to the axilla; and is not heard in the back. It is not, necessarily, associated with venous pulsation.

Mitral Obstruction and its Consequences.—Dr. E. R. Bruch,† at a meeting of the Pathological Society of Philadelphia, exhibited a heart the mitral opening of which was nearly occluded by an epigglotic-shaped enlargement

---

* *N. Y. Med. Jour.*, March 3, 1883.
† Ibid.

of one of its leaflets, permitting during life reflux. The result was dilatation and hypertrophy of the left auricle to twice its proper size. The right ventricle is very much dilated, the walls of less than half their proper thickness, and the capacity of its cavity doubled. The tricuspid valve was insufficient on account of this dilatation. There was during life a broad area of dulness, extending to the right of the median line. The murmur heard was both systolic and presystolic. The second sound was much accentuated. The first sound was distinct over the ventricle, but obscured by the murmur at the apex. The immediate cause of death was pulmonary repletion with blood which induced right heart failure, as in cases of aortic obstruction it is induced by left ventricle failure.

Pleurisy with Heart Disease.—Dr. Bucquoy* has observed a number of cases of pleurisy occurring in the course of heart disease. He says that though the patient may be weak from this disease of the heart often associated with albuminuria, the inflammation has all the characteristics of subacute idiopathic pleurisy, and is entirely distinct from hydrothorax. It usually pursues a favorable course, ending at length in absorption. He would hasten a cure, and if absorption is tardy recommends tapping, especially if there is dyspnœa or other urgent cardiac symptoms.

Mitral Stenosis—Prognosis.—Dr. Dureoziez † finds that mitral stenosis, simple or combined with other valvular diseases, permits, in exceptional cases, the life to pass the sixtieth year. The complication of mitral regurgitation does not aggravate the prognosis. The co-existence of aortic insufficiency renders it more grave. The complication of tricuspid lesion is of very serious import.

---

* The *Med. Rec.*, March 10, 1883.
† Ibid, Oct. 27, 1883.

Aortic Regurgitation Makes the Pulmonary Second Sound Inaudible.—Every auscultator has noticed that the sound of aortic regurgitation may be pretty loud at the base of the heart, and at the same time be nearly or quite inaudible at the apex. This is not true of the louder class of murmurs, but is true of those which are loud enough to obscure or wholly cover the pulmonary second sound. It is possible, in such case, to find the pulmonary second sound audible, more or less clearly, at the apex, while the regurgitant sound will be conducted down the sternal bone much more loudly than down the heart tissues. Dr. Begbie has called attention to these facts.

Stenosis of the Right Auriculo-Ventricular Opening.— Dr. Kucher reports the cure of a child twenty-four hours old in whom the right auricle was larger than the left, the borders of the thin curtains of the tricuspid valve were sealed together, and made a funnel with two openings in the bottom of the funnel that would admit a Simpson's sound. The material of the valve was thickened and hard. This he ascribes to fœtal endocarditis confined to this valve.

He also found in the left auriculo-ventricular opening a transverse chorda tendinea, and says that such a chord was recognized during life by Schroetter by a peculiar musical note.

There was nothing striking in the ante-mortem history except that the chest arched as in rachitis, and that the child was bluish.

At the end of twenty-four hours the child became cold, and the doctor was sent for. On his arrival it was dead.

Dr. Kucher says that while fœtal endocarditis is not uncommon, tricuspid stenosis is very rare. Rauchfuss found only two similar cases on record; one reported by Peacock and one by Romberg. In Romberg's case, a boy of four years, the opening was very narrow, and the tricuspid valve had disappeared. In Peacock's case, a girl of seven months, the stenosis was not considerable, and was apparently due to the thickened valves with inflammatory deposits on their outer walls. There were two defects in the intraventricular septum.

Tricuspid and Mitral Stenosis.—At a meeting of the London Pathological Society Dr. Bedford Fenwick[*] showed a specimen of tricuspid stenosis from a woman aged thirty, who had rheumatic fever at fifteen and afterwards suffered from winter-cough and dyspnœa. There was marked distention of the jugular veins, but no cyanosis, cardiac dulness extending far to the right, a well-marked presystolic apex thrill, and systolic and presystolic apex murmur, to the right another systolic and presystolic murmur. At the post-mortem examination both auricles, but particularly the right, were very much dilated, the ventricles small. The tricuspid and mitral valves were greatly thickened, shortened and agglutinated. Cases of this kind are now known not to be very rare. Since his table of forty-six cases he had been able to collect twenty-three more, twenty of whom were in females, average age 31.7 years. In every case the mitral valve was more changed than the tricuspid, and in all the general health had been good. The great dilatation of the right auricle caused increase of dulness to the right, and afforded a means of diagnosis.

Dr. Vohsen[†] finds that in nearly one half of the children suffering from acute articular rheumatism endocarditis occurs and results in marked valvular insufficiency. The mitral valve was most frequently attacked. Endocarditis appears usually in the first or second week.

---

[*] The *Medical Record*, Jan. 20, 1883.
[†] Ibid.

Pericarditis is also a frequent complication. Salicylate of soda, while it exerts a favorable influence on the joint disease, seems not to prevent the cardiac complications. The mild forms of articular rheumatism are especially liable to be followed by cardiac disease.

Dr. O'Hara reported to the Pathological Society of Philadelphia,* the case of a patient, a laborer thirty years of age. At first the usual symptoms of cardiac disease. On admission, with the forearms flexed to a right angle, the brachial arteries became prominent with each impulse of the heart; the pulsations of the carotids were wavy and prolonged; the temporals were tortuous and pulsated visibly; no retinal arterial pulsation was seen. Retinal venous pulse was marked, but venous pulse was found elsewhere. In the second left intercostal space a systolic impulse was observed. The pulse struck the finger with considerable force, but at once lost most of its volume. This was exaggerated by raising the hands above the head. No hepatic pulsation was felt over the second right costosternal articulation, the closure of the aortic valves was distinctly heard and a slight diastolic murmur. A systolic murmur was also heard over the same spot. The systolic murmur was nearly lost in the carotid and sub-clavian arteries.

The aortic valves were insufficient and thickened; the posterior leaflet was normal in shape, but the others were curled on themselves on the aortic side. The stenosis was slight. The mitral orifice was buttonhole shape, and the valves failed to close on account of calcareous deposit in their tissue, extending into the auricle. Attached to the valve a bony substance hung into the ventricle one eighth of an inch in diameter. The left auricular appendix was much hypertrophied. The valves of the

---

* *N. Y. Med. Jour.*, Jan. 20, 1883.

tricuspid and pulmonary orifices were normal. The pulmonary artery was considerably dilated. Dr. Eskridge thought that the demonstration of the left auricular pulsation was important, as Dr. Broadbent had so recently advanced an opposite view: (Was there a functional murmur in the pulmonary opening due to enlargement of the artery?)

Dr. Bruch said that he would like to go on record among those who had observed auricular pulsation in cases of mitral obstruction, in which the stenosis was extreme.

Trumpet Bruit.—Dr. Eskridge said that Hayden* referred to bony deposits in the aorta and its valve as follows: Sir Dominic Corrigan exhibited before the Pathological Society of Dublin (Proceedings, vol: ii., new series, Feb., 1864) the heart of a young woman, in which the root of the aorta had undergone osteoid transformation; it was likewise greatly dilated, and the aortic valves had been rendered thereby inadequate. During the patient's last illness a systolic murmur of metallic quality, appropriately designated a 'trumpet bruit,' was audible at the base and in the ascending aorta and carotid arteries. There was likewise a soft diastolic murmur. He regards a "trumpet bruit" as absolutely diagnostic of bony deposit in the aorta, either in the form of a "rim of bone," of a projection, or "tongue of bone." In the same paragraph Corrigan refers to Dr. Banks' specimen of a tongue of bone projecting into the aortic orifice.

Lying-in Murmurs of the Heart.—Dr. Angel Moncy† publishes a paper on this subject, the chief conclusions of which are the following. One kind is endocardial-like, and may be heard over any part of the precordia. The

---

* *Diseases of the Heart and Aorta.* Vol. ii., p. 839.
† *Medico-Chirurgical Transactions.* Vol. 65, 1882.

second is friction-like and not conducted just above and to the left of the xiphoid cartilage. The third is very loud, of *curious* quality, not conducted, and very capricious. Some of these murmurs occur in seventy-five per cent of the lying-in, and a large part of them are heard over the right heart. These murmurs are for the most part functional murmurs and need not excite any alarm.

Relative Stenosis of the Cardiac Orifices.—A theoretical explanation of the transient heart murmurs has been advanced.* In the normal heart, the size of the orifice multiplied by the velocity of the blood-current must equal the cubic contents of the heart cavity. If the latter be greater, we have stenosis. Hitherto but one factor has been taken into account, viz., the diameter of the orifice, but it is clear that if with a normal exit the capacity of the heart be increased, the former becomes relatively too small, and we have a relative stenosis. The auriculo-ventricular opening partakes in the dilatation of the cavities, to the extent indeed that the unchanged valves become too small, and insufficiency results, But the arterial orifices do not dilate as the ventricles enlarge and hence become too small. As regards the third factor— the velocity of the blood-current—a certain relation exists between the capacity of the heart and the size of its orifices, which permits changes in the velocity of the current within certain limits. But when through congenital defect or in consequence of pathological changes the ventricle can barely empty itself, a slight increase in the velocity will produce a murmur. On the other hand, a pre-existing murmur may disappear on the retardation of the blood current. Witness the disappearance of a systolic murmur after a severe hemorrhage. We should make a distinction between organic and functional or clinical stenosis. Organic valvular changes are sometimes

---

* The *Med. Record*, Sept. 23, 1882.

found after death that gave rise to no murmurs during life. In acute endocarditis or during a high fever, there may be a weakening of the muscular walls of the heart and consequent dilatation. Here the relation between the ventricle and its orifice is altered, and the latter becomes stenosed. The sound produced is a blowing murmur, heard oftenest and loudest over the left heart, and of course systolic. After copious hemorrhage loud blowing systolic murmurs arise, often not till the vessels are refilled by the absorption of the water of the tissues.

From insufficient nutrition, the muscular tone of the heart is lowered and there is consequent dilatation. When this condition exists, the orifice being relatively too small for the ventricle, only a part of the contents of the latter can pass out; the balance recoiling against the advancing blood-stream causes a murmur.

Before proceeding with the consideration of the treatment of valvular disease, I append a letter, in which my opinion is asked regarding the relation between heart and kidney lesions, and the reply, in which are set forth the views I have entertained upon this subject for many years.

SPRINGFIELD, VT., Aug. 20, 1883.

Prof. Alonzo Clark, M.D.

DEAR SIR: The following case reported at a local medical society excited some interest among the members, and it would be gratifying to them to know your opinion of it. Is it possible, do you think, to decide which is the initial lesion when the heart and kidneys are involved (as in this case) by the clinical history, and by the stage of development of each lesion, as shown by the autopsy? If it would not encroach too much upon your time, would you have the kindness to give us your opinion? With high esteem,

I am, very respectfully, yours,
DANIEL W. HAZELTON.

VALVULAR DISEASE.

Case.—Nov. 1880. J. B., male; age 72; married; farmer by occupation, and a man of good habits. When first seen he gave a history of repeated attacks of acute articular rheumatism during the preceding 20 years. For a long time he had had dyspnœa, which was made worse by exertion. The dyspnœa persisted and was more marked than ever before. For days he was unable to maintain any but the sitting posture. Pulse was intermittent. There was a cough and expectoration occasionally streaked by blood. There was fulness of the superficial veins of the neck, and distinct pulsation of the jugulars. Lower extremities were œdematœus, auscultatation gave a systolic murmur heard best at the apex of heart and indistinctly at angle of scapula. Percussion showed the area of cardiac dulness to be increased, extending from $\frac{1}{2}$ to 1 inch farther to right of sternum than normal. Urine on first examination and for some time subsequently was about natural in quantity and color; specific gravity about 1015, and contained a truce of albumen. At different times during the succeeding year the patient had small hemorrhages from the lungs, and at various times the lungs gave dulness on percussion, when dyspnœa was increased. He also had hemorrhages from the bowels, one of which was profuse. Patient died with symptoms of uræmia and pericarditis December, 1882. Autopsy: Pericardium congested and bound by recent adhesions to heart. Heart was moderately dilated. Right side more dilated than left. Left side about normal—weighed 22 oz. Mitral valve had an abundant deposit on it of calcareous matter, also the margins of the auriculo-ventricular orifice.

The kidneys were slightly atrophied. Their capsules were easily removed. One cyst about the size of a pea was found between the capsule and cortex of one kidney. They were of about the natural reddish color, but were studded with very fine and rather indistinct millet seed granulations.

On section, the cortex and pyramids, presented their natural color and structure so far as the unaided eye could determine. The cortices were not diminished in breadth. Other viscera were normal.

"DEAR DOCTOR :

"It has been the subject of much discussion, when cardiac and renal disease are found in the same patient, which preceded. This question, in my mind, divides itself into two, and would be answered as follows: When it is the *large* kidney, with albumen and casts in the urine and much œdema during life, then in nine-tenths of the cases the heart disease is primary and the kidney affection is secondary and consequent. When, on the other hand, the kidney is *contracted*, and during life the albumen is scanty and often absent, the casts are uncertain, there is but little œdema, and there is no *distressing* valvular disease, then the chances are at least even that the renal disease is primary.

"In your case the articular rheumatism had been recurring for twenty years, and " for a long time the patient had had dyspnœa." How long ? This question is important. If he was pretty well and could attend to his business for six months or for some years after the rheumatic attack or attacks, and then the difficulty of breathing came slowly, with perhaps at times an urgent attack, and if especially this difficulty increased rapidly after the œdema appeared, then you have the history of rheumatic mitral disease as the primary affection and the renal is secondary.

"In your case there was calcareous deposit on the mitral valve and in the margin of the orifice. That means years.

"Then, again, "the left side of the heart was about normal," the right was dilated, and the heart weighed 22 oz. It is evident, then, that there was hypertrophy somewhere· It was not in the left side. There must have been hypertrophy with the dilatation of the right side. This should be looked for because of the long standing stenosis of the mitral valve. You do not state that there was a diastolic mitral murmur. Yet, from the hypertrophy of the right ventricle and the condition of the mitral valve and orifice, I infer that it did exist at one time and had been lost as the strength of the heart diminished toward the end of life.

" Then as to the kidneys. They do not appear to have been in a state of advanced disease. They were '"slightly atrophied." They were finely granulated, but their capsules were not unnaturally adherent, while the internal structures were not perceptibly changed. There was one serous cyst, but only one, and that was small. This appears to be the early stage of the contracted kidney, a disease which sometimes reduces the weight of the organ to one and a half ounces.

" When the small kidney produces disease of the heart it is the left side that is affected and hypertrophy is associated with it, or it is hypertrophy alone. In my judgment, then, the chronic disease of heart antedates the disease of the kidney, but probably was not the cause of it. The contracted kidney is not that which usually follows cardiac disease, but the large one. How a kidney of the kind you have described should have occurred in such a case is a problem I cannot solve. If it was really produced by the heart it is a noteworthy exception to the rule I have referred to. If it was the result of the unknown agencies which usually produce the contracted kidney, even then the concurrence is worth remembering. One other supposition is possible. The influences that produce the contracted kidney may have been operative before the dyspnœa was experienced and when that came, the congestive state of the kidneys, which comes with it, may have counteracted the contraction and suspended it. Whether this is possible I do not know. If it is, then whether it explains this case can be better judged of when we know for how many of the twenty years the dyspnœa existed. The duration of the cirrhotic kidney may never be known, but after it is recognizable the patient often lives three or four years. Still it is difficult to suppose that it can have a total duration of twenty years.

"There is one other point in the case. Your patient had pericarditis at the end. You state that the urine contained a little albumen, and your idea that he was also uræmic is altogether probable. This state of the blood is quite capable of producing pericarditis and probably did in your case.

"The other points in your case are easily explained. The cough, the expectoration, and the blood stains were produced by the distention of the vessels consequent on the mitral regurgitation and stenosis. The intestinal hemorrhage from the overloaded state of the right heart and its unwillingness to receive the blood of the hepatic and portal vessels and the consequent rupture of some intestinal veins. The fulness of the veins in the neck and the pulsation in the jugulars were due to regurgitation through the tricuspid opening, for while the right ventricle was dilated its auriculo-ventricular opening would be enlarged also, and so render the valve insufficient to guard it.

"You say 'at various times the lungs gave dulness on percussion,' but you do not state in what part of the chest this dulness was found, or whether it was on both sides. If it was on both sides and in the lower part of it, it was dependent on double pleural serous effusion—an event not uncommon in cardio-renal disease. It is a part of the dropsy not caused by inflammation, and it is therefore painless, but it greatly embarrasses the respiration. It is possible for congestion of the vessels of the lungs to so far exclude the air from the air cells as to make dulness. This, is however, somewhat peripatetic, moving from place to place in the lung or from one lung to the other and is apt to be attended by hæmoptysis.

"Finally, you appear to have found no fluid in the pericardium only lymphy adhesions, yet you found percussive dulness on the right of the right edge of the sternum one half to an inch in extent. This was probably due to a serous effusion in the pericardium that was absorbed before death."

LECTURE XIII.

PROGNOSIS AND TREATMENT OF VALVULAR DISEASE.

In regard to the treatment of valvular disease, it is so entirely bound up in the subject of treatment of cardiac diseases in general that it does not require attention at this time. The question will naturally occur to you, how much havoc can be made by these diseased valves? How dangerous is it? The reply would be this: Obstruction at the aortic opening is the least dangerous of all, because of the compensation that can be furnished by hypertrophy of the heart. Obstruction at the mitral opening will be grave or not, depending upon its extent. As I told you, one day it may become a very grave matter, producing cyanosis that may be visible in the face, a congestion of all the important organs of the body, and leading to rather a slow death, that is, an illness that may be protracted through several months, or half a year, or even a year, and finally terminating in death. But diseases of the mitral valve usually have a slow ending through affections of the kidneys and dropsy, particularly œdema. Regurgitation at the mitral valve goes in the same class; it stands with obstruction of the mitral valve in regard to its mode of termination; that form of disease that is most likely to have a sudden termination, that condition in which the patients are sometimes told they may drop down dead in a moment; and this is a very unwise thing to tell a patient; it makes him anxious all his life, fearing at every moment he may drop down dead. The condition that is most likely to be attended by this sudden result is regurgitation at the aortic opening, by a return of blood into the ventricle after it has once entered the aorta. You can easily see that in this case, depending upon how much regurgitation there is, there will be a diminished quantity of blood circulating in the arteries of the body. The brain will not receive its due amount, and syncope may take place, you cannot say when. It does not require a very grave lesion of this kind to lead, in certain instances, to sudden death, and it almost always occurs by a sort of syncope, the brain receiving too little blood to stimulate it to proper action, and to react upon the heart. The heart ceases to beat and the patient is dead. And yet do not give these patients unnecessary alarm. They may carry this lesion that leads to regurgitation from the aorta into the ventricle for a great many years. The number of instances in which persons die with this lesion suddenly, compared with those who die a lingering death, is very small. It is an old error to suppose that persons who have cardiac disease must die suddenly. They may, and others may. A good many years ago, when I was just entering upon the profession, I took up the study of the cause of sudden deaths; and one of the first that occurred which I studied was of a young woman in Bellevue Hospital. She had had typhus fever. It was in 1847. I was not so young in the profession as I was thinking. Well, typhus fever was prevailing there to a remarkable extent. She had got through with her typhus fever, and was in a condition of convalescence. She was able to go about the wards and take care of other patients to a certain extent, but she felt a little weary at times. After eating dinner she had lain down upon her bed and taken a nap. Another patient, her next neighbor, had done the same thing. She woke from her nap, raised herself in bed, and said to her neighbor, "I have not felt so well as I do at this moment since I was first taken sick," and had scarcely finished the sentence when she fell back senseless, dead. At post-mortem examination we found there was an effusion of blood at the base of the throat, involving all the nerves of respiration, compressing them to a certain extent, so that the respiration stopped suddenly. In following up the inquiry, I found within six months six cases of the same kind. Since then I have not seen one. Sudden death, then, has a variety of causes. In a person who is known to have cardiac disease it is usually ascribed to that affection, but it is not a matter of course that it will be so. And then you are to bear in mind that in certain forms of disease of the heart there is greater danger at the brain than there is at the heart itself. That is, hypertrophy of the heart without obstruction is a rather frequent cause of apoplexy. An hypertrophy of the heart without obstruction between the arteries of the brain and the heart itself is apt to lead to apoplexy.

In Regard to Remedies.—In cases of dizziness from heart disease, See* asserts that iron is absolutely indicated, though in many cases the iodide of potass gives excellent results.

Anginous attacks should of course be combatted by hypodermics of morphine, the administration of chloral, or the inhalation of nitrite of amyl during the access, and by the bromide of potassium and digitalis or convallaria during the intervals.

Excellent results are claimed in the treatment of cardiac dropsy by caffein. Begin with a dose of 7 grains, gradually increasing to 15, 30, and even 45 grains per diem. Such large doses, however, often cause severe pain in the stomach.

Adonis vernalis† has long been used as a dropsy

---

* *Med. Rec.* Sept. 22, 1883.
† The *Medical Record*, Oct. 6, 1883.

remedy in Southern Russia, and Prof. Bolkin has employed it extensively, and Dr. Budnoff reports the result. It was found to be of value only in the dropsies that were preceded by cardiac disease. The heart-beats were much strengthened. The size of the organ was diminished, and its tones were much louder. The systolic murmur of aortic stenosis especially was intensified. The heart's rhythm became more regular, and the pulse slower and fuller. The daily excretion of urine was increased greatly, sometimes rising from ten or twelve to eighty or ninety, ounces a day. The subsidence of œdema was proportioned to the flow of urine. The medicine was given in infusion of the strength ℨ i. to ℨ vi., to which two drops of the oil of peppermint was added. The dose of this was a tablespoonful every two hours.

Dr. Bubnow* of St. Petersburg, says that it is a popular remedy in Russia for dropsy. Under its influence the cardiac contractions are increased in force and diminished in frequency, and the urine was augmented, and no longer contained casts or albumen. Experiments on both cold and warm-blooded animals with different preparations (infusion, aqueous and alcoholic extracts), showed that its action is to stimulate both the motor ganglia, and the imbibitory apparatus of the heart, and to raise arterial tension. Dr. Bubnow prefers adonis to digitalis, and says that, like convallaria, it has no cumulative action. A glucoside *adonidin* containing the active principle has been isolated. The effects of adonidin are very similar to those of digitalin. It is not improbable that the same similarity may be found to exist between adonis and digitalis as exists between hyoscyamus and belladonna.

The therapeutic uses of convallaria are thus stated by Prof. See† :

---

* N. Y. *Med. Jour.*, June 16, 1883.
† *Am. Jour. of Med. Sci.*, Oct., 1882.

1. In palpitation due to an exhaustion of the vagus nerve or paralytic palpitation.

2. In simple want of rhythm with or without hypertrophy, with or without lesions of orifices or valves.

3. In mitral stenosis the contractile power is increased as shown by the sphygmograph.

4. In insufficiency of the mitral valve, particularly when there is passive congestion of the lungs and consequent dyspnœa.

5. In aortic regurgitation (Corrigan's disease) the good effects are shown in the pulse and easier respiration. It gives strength and energy to the heart.

6. In dilatation, with or without hypertrophy; with or without fatty degeneration; with or without sclerosis of the muscular tissue.

7. In all cardiac affections giving rise to dropsy and anasarca surpassing all other drugs.

The combination of convallaria with the iodide of potassium is most valuable in cardiac asthma.

In the *Medical Record* for Feb. 24, 1883, may be found the report of twenty-one cases treated with this drug, all heart disease in some form; the ages of the patients from eleven to seventy years.

The preparation was an infusion ℥ ½ to ℥ i. to a pint, in tablespoonful doses every two hours. In seventeen cases there were absolutely no results, and in the others the improvement was trifling. The writer therefore does not admit that convallaria is an efficient substitute for digitalis. He does admit, however, that the specimens of convallaria used by him may have been of a quality inferior to that used by other observers, who have reported such brilliant results.

Dr. B. Robinson[*] exhibited to the N. Y. Pathological

---

[*] The *Medical Record*, Jan. 13, 1883,

Society the hearts and kidneys of two persons who were similarly affected, and in whom both digitalis and convallaria were equally ineffectual in producing diuresis, in both of whom the contracted kidney was found, regarding which Dr. Robinson expresses the following opinion: "We should be prepared to assume that with a small quantity of urine, with nothing specially abnormal about the specific gravity, and failure on the part of the kidneys to respond to cardiac stimulants, as in the case recited, we might reasonably expect to find that form of kidney degeneration."

Dr. A. A. Smith's* experience with convallaria had led him to think it a good diuretic. It had seemed to him that in cardiac hypertrophy the remedy was not indicated. In chronic renal disease, with chronic hypertrophy, according to his observations, it aggravated the symptoms.

In cases of enlargement of the heart in which dilatation predominates it serves a very good purpose. He was not however, prepared to give up digitalis, and substitute convallaria. Nor was he quite sure the experiments of Ott had not demonstrated that it does not act through the pneumogastric.

Dr. Vanderpoel† found in a case in which there was mitral stenosis and insufficiency with irregular action of the heart and œdema, that eight minim doses of the fluid extract of convallaria twice a day brought marked improvement, removed the œdema in ten days, and brought a more regular heart action.

Dr. Delafield said it made the heart action more regular and slower in certain cases, and in general condition the patient improved very much. He had found, like Drs. Vanderpoel and Barker, that there was a great difference

---

* *Medical Record* (N. Y. Acad. of Med.), April 14, 1883.
† N. Y. *Med. Jour.*, Apr. 14, 1883. (N. Y. Med. and Surg. Soc.)

in patients as to dose. For some five drops of fluid extract every three hours was enough, while other patients required drachm doses. He thinks he saved the life of a woman, over seventy years of age, overwhelmed by pneumonia, with twenty-drop doses of the fluid extract every three hours.

Dr. Hurd* has translated Prof. Germain See's opinions of this new medicine, which are mostly embraced in the following extracts : In the form of aqueous extract of the entire plant, given in doses of half a gramme to a gramme and a half daily, it produces on the heart blood-vessels and respiratory organs, effects constant and constantly favorable, viz.: a slowing of the beatings of the heart, with often a restoration of the normal rhythm, augmentation of the energy, also of arterial pressure, the respiratory force is increased. . . . In all cardiac affections indifferently, from the moment the watery infiltrations appear, it has an action prompt and certain. Dr. Hurd has tried the fluid extract in doses of five drops every four hours in two cases with very satisfactory results.

Dr. Troitsky † says it may be used to make the heart's contractions stronger, slower, and more regular, especially in cases of organic disease, or those in which increased frequency depends on changes in the nerve centres; to lower the temperature so far as its influence on the heart may permit; to increase arterial tension; to increase the secretion of urine in dropsy; to lower reflex action. He says that it acts better in mitral regurgitation than in the diseases of the aortic valve and mitral stenosis. In this respect it is like digitalis—ten or twelve grains of the flowers to six ounces of water, three or four tablespoonfuls a day. Convallaria in large doses does not

---

* The *Medical Record*, Sept. 9, 1882.
† N. Y. *Med. Abstract*, April, 1883.

stimulate the cardiac terminations of the vagi—does not stimulate the cardiac nerves. It stimulates the central inhibitory apparatus. It paralyzes the motor centres situated within the heart itself. It at first accelerates then retards the respiration, and finally stops it just after the heart has stopped, and he thinks, because the respiratory centres are paralyzed by the venosity of the blood. Temperature at first rises half to one degree C., then falls considerably because it paralyzes the vaso-motor centres. It first stimulates then paralyzes the vaso-motor centres, hence increased and then diminished vascular tension. On the brain it produces anæmia and sleep; applied directly to striated muscle the extract produces complete loss of contractility.

The trial of this drug in the Roosevelt Hospital, under the direction of Dr. Delafield, reported by Dr. Taylor,* is summed up as follows: "Of the five cases of cardiac disease four improved—one very markedly—after digitalis had failed; one improved at first, then grew very much worse and recovered on digitalis. One was hopeless. Case XI., already referred to, was as brilliant an example of the benefit to be derived from a drug in certain instances as I saw during my hospital experience." Fifteen other cases are reported in which the medicine was used in other diseases. One, a case of pneumonia, hardly a less brilliant illustration of the power of the medicine than the Case XI. of heart disease. The fluid extract was given X minims three times a day to XX minims every four hours.

Dr. Polk† reported to the Practitioners' Society a case of palpitation in which the heart would beat 240 times in the minute, in one instance for three days. Drs. Polk,

---

*The *Medical Record*, Feb. 13, 1883, and a previous number.
† Ibid, Feb. 3, 1883.

Peabody, Flint, and Metcalfe had all failed to find any organic change in the heart. The patient was a young man of twenty years. In one of these attacks Dr. Polk gave him ten minims of the fluid extract of convallaria hypodermically, and one half hour later repeated it. Five to ten minutes after this the pulse fell to 120, and fell not gradually, but suddenly. Fifteen-drop doses every three hours. The next morning the pulse was 90 in the minute. The intervals were lengthened to four hours, and the next day the medicine was suspended. There was no return of palpitation.

Dr. Beverly Robinson* reported to the Practitioners' Society an eleven years old asthma in a man sixty-three years old, inherited, which was greatly relieved by the fluid extract of the root of convallaria. The first dose of fifteen minims gave sensible relief. Then it was four minims in eight of syrup of tolu and water sufficient to make a drachm, every three hours, then five drops of the extract every three hours. It was continued through December, having been begun Oct. 4, 1883.

Dr. Ott finds that the slowing of the heart by digitalis is due to a cardiac inhibitory excitation; with convallaria some other part of the heart is acted on (he thinks the muscle). Digitalis, as a rule, does not primarily accelerate the heart; convallaria does. After section of the spinal cord digitalis is powerless to increase arterial tension, while convallaria is not. If we compare the action of the drug with other cardiac agents, as aconite, or astragalus molissimus, it is found that it does not belong to this group. As these resemble each other in their action, yet many important differences exist. So does convallaria differ from digitalis in several important

---

* The *Medical Record*, Feb. 3, 1883.

particulars. The great rise of arterial tension would indicate its value in dropsies. It is a drug which must not be pushed to any great extent for fear of producing spasm.

Salicyl Compounds in Rheumatism.— Dr. Maclagen * again brought forward his views. He urges that small doses are ineffectual, but that we employ large doses and at frequent intervals—say twenty to forty grains every hour for six hours, or until there is relief from pain. Dr. Maclagen thinks that we may, by the early and free use of salicin, prevent valvular inflammation. Dr. Sansom, while indorsing much that Dr. Maclagen has said, was not so sanguine as to the preventive action of salicin.

Cactus Grandiflorus in Heart Affections. — Dr. M. O'Hara† at a meeting of the Philadelphia County Medical Society said that his attention was first called to this cactus some time ago, and failing with the usual remedies in a case of mitral disease with œdema, dyspnœa, etc., he gave five minims of the fluid extract of the cactus grandiflorus. Every symptom improved, and in two weeks the patient was quite another man. He had two other similar cases which give similar evidences of the virtues of the cactus. He says that if he can trust a limited experience the cactus is a pure heart tonic.

He refers to various other remedies recommended by different persons who think that their action is similar to that of digitalis, but he has not tried them.

Dr. Finny‡ thinks that too much dependence may be placed on the physical signs of mitral regurgitation as evidence of organic disease; that such signs may be due to purely functional derangements and weakness of the

---

* The *Medical Record*, April 7, 1883.
† N. Y. *Medical Journal*, Nov. 17, 1883.
‡ The *Medical Record*, March 17, 1883.

heart, or to an altered condition of the blood; that the blood murmurs in the heart and large vessels may be louder than those produced by valvular lesions; that the danger of valvular disease is enormously increased by weakness of the cardiac walls; that the lowering treatment of the heart's force is rarely if ever required in heart disease; that indications for treatment in diseases of the heart should be sought from the evidences of the condition of the heart muscle, and not from those of the condition of the valves.

Dr. Leech* said his general plan in the treatment of cardiac dropsy was to place his patient in bed, secure rest for some time, and then administer diuretics. When diuretics failed, purgatives rarely succeeded. He had not seen benefit from the use of the hot air bath. He had entirely failed to see any benefit from the use of pilocarpin. Unlike tapping in cirrhosis, he had seen but little loss of strength in tapping for cardiac dropsy. He had but once had peritonitis follow tapping. Southey's capillary tubes for draining the legs in general anasarca were of great value, although he had seen phlegmonous erysipelas follow their use in one case. He had found that diuretics would act often after rest and good diet, though they had previously failed.

Dr. Allbutt had found diaphoretics very uncertain. He regarded hydragogues as very useful. The air bath was unsuccessful. He hesitated to tap the patient's leg in fear of erysipelas, but otherwise the operation was almost brilliantly successful. He used Southey's tubes with satisfaction. He delayed tapping on account of the exhaustion it produced.

Dr. Thomas had found diaphoretics inactive, but some-

---

* British Med. Association, Aug., 1882; *Med. Record*, Aug. 26, 1882.

times they were satisfactory after purgatives, but generally no real benefit from attempts to act on the skin.

Dr. Carter had used the hot air baths and hot drinks with marked benefit, especially in dropsy after scarlet fever, and had frequently seen diuresis follow. For cardiac and other dropsies digitalis had in his hands proved to be exceedingly useful. He had resorted to early tapping, but he had used a very small trocar, which would allow only a slow removal of the fluid, believing that thereby the liability to exhaustion was considerably diminished. He referred to three cases. The *first* was ascites from cirrhosis, and tapping was followed by gradual exhaustion. The *second* was ascites with Bright's disease. In this tapping, doubled the quantity of urine and was followed by relief. The *third* was ascites with cardiac disease, and each tapping was followed by diminution of the urine one half, although relief was afforded. The patient was tapped seventeen times. The patient with cirrhosis died, the other two left the hospital in a comparatively comfortable condition. He had used Southey's tubes for the legs with good results. He believed the use of hydragogues should not be constant, but occasional. He preferred rhubarb, jalap, etc., with elaterium. Dr. Roden recommended iodide of potassium combined with full doses of nitric acid. Some preparations of iron with saline cathartics possessed special advantages.

Dr. Phillip's had used Southey's tubes successfully, but regarded hypodermic injections of pilocarpin as one of the most valuable agents that could be used.

LECTURE XIV.

ANGINA PECTORIS

Is pretty plainly a neuralgia. Whether the change that produces it is in the muscle of heart, caused by scanty supply of blood to the organ, or is in the terminal branches of the cardiac nerves, or as the *Medical Record** expresses it, referring to Dr. Allen Sturge's opinions, due to "a commotion spontaneously developed in the gray substance of the sympathetic ganglia of the cardiac plexus, this being transmitted to the spinal cord and brain; a commotion spontaneously developed in the cervical sympathetic ganglia, which give off branches to the cardiac plexus, or in the ganglia of the vagus; a spontaneous commotion in the part of the gray substance of the cerebrum which may receive impulses coming from above and below; or to a spontaneous commotion in the parts of the gray substance of the cord in communication with these ganglia, by means of bands of nervous substance passing from the cord to the great sympathetic."

Angina arising from tobacco is a rare thing. Tobacco causes intermittence, arythmia, etc. When it causes angina it is due to the slow action of the nicotine on the coronary arteries.

Angina arising from alcohol signifies endarteritis of the coronary arteries as well as degenerations or sclerosis of the myocardium. The mischief is done, the suppression will not avail.

Gouty angina of the Germans is a cardio-vascular

---

* *Med. Rec.*, Sept. 22, 1883.

lesion and resists treatment; it is like alcoholic angina caused by endarteritis; the alkalies can do little for gout of the heart.

Angina of organic origin, alterations of the coronary arteries, degeneration of the cardiac muscle, dilatation of the cavities, and lesions of the aorta, oftener cause attacks than mitral lesions.

Hysterical angina is very rare. These neurotic anginas are nearly as dangerous as those of organic origin.

There is, in fact, no paralysis of the sympathetic in angina pectoris. The disease is in reality attended by excitation of the cardiac sympathetic nerves and coronary vessels; the latter being in a state of erethism, there is no paralysis in the case; but there is not even excitation of the sympathetic nerves, accompanied by the contraction of the blood-vessels in general, the disease, so far as nerves are concerned, being partial and limited. If it were possible to galvanize the cardiac sympathetic nerves, would you not augment rather than diminish the vasomotor contractility?

Obstruction and Regurgitant Mitral Murmur with Healthy Valve; also Angina Pectoris.—A case is reported in the *Lancet*, Jan. 27, 1883, in which the patient died in a paroxysm of angina pectoris, left hypertrophy and dilatation and the mouths of the coronary arteries obstructed. The writer believes that angina pectoris is caused by ischæmia of the heart.

The patient had a presystolic and a systolic murmur, yet the mitral valve was normal by the water test and was perfectly healthy in appearance. The presystolic murmur had in this case its usual distinctive feature—a rough vibrating quality resembling in this respect the sound produced by the lips when thrown into vibrations by the expired breath. The writer calls it a blubbering sound.

The author explains the murmurs as follows: "I do

not propose to discuss the various explanations of the production of a mitral systolic murmur without mitral lesion. Suffice it to say that in this case, taking into view the considerable dilatation of the left ventricle, it seems to me probable that, although the ventricle held water when filled from an opening made through the apex, the ventricular contraction during life occasioned some regurgitation. A regurgitant stream, be it ever so small, may give rise to a murmur. I submit this explanation without undertaking at this time to account for the apparent incongruity that the mitral may admit of some regurgitation during life and yet be found competent after death."

Hysterical Angina Pectoris.—Marie* reports two cases in hysterical patients. In one the pain commenced in the left little finger, ran up the arm to the breast of that side. During the attack, which often lasted some hours, the pulse of the left radial artery became imperceptible, the lower parts and the whole left side became cold. In the second case the attack was often preceded by general malaise; then a sudden violent pain in the præcordial region; accompanied by extreme anguish, invincible terror. The pain then extended towards the neck and left arm and little finger, and sometimes toward the leg. The attack lasted ten to twenty minutes, and the face, which was at first pale and cold, became red and warmer. He compares them to a form of angina called vaso-motor by the Germans, but does not say whether there was heart disease.

Prof. See's Treatment of the Paroxysm.—See has abandoned the use of chloral. It induces sleep, but has no effect on the circulation.

Hypodermic injection of morphine, a centigramme

---

* The *Med. Rec.*, June 2, 1883.

of the hydrochlorate, one sixth of a grain, may be repeated, may be given daily to prevent return.

Nitrite of amyl is one of the most active and useful means, three or four drops inhaled from the open palm. In the first period of the action the vascular tension is lowered, the vessels are dilated, the action of the heart is very much quickened, the respiration is more free and easy. At a more advanced period, the pressure remaining lowered, the heart and respiration become slowed. There is sanguineous irrigation of the coronary arteries, as in other vessels, the cardiac ischæmia ceases. From more than the dose indicated there is risk of cardiac syncope. The patient soon becomes accustomed to the remedy, so that its good effects become less and less.

Nitro-Glycerine.—The effects of this are very much like those of the nitrite of amyl, only slower. The dose is one or two drops of a one per cent solution.

See's Treatment in the Intervals.—Bromide of Potassium. Under its action the patient becomes less impressionable to physical and psychical influences which might provoke a paroxysm, but it produces debility, which is more or less permanent.

Digitalis. When the attacks result from cardiac atony or degeneration it is better than the bromide; it sustains the action of the heart and is every way preferable.

Of electricity he speaks very doubtfully; it may even kill, as in Duchenne's case.

Angina Pectoris and Counter Irritation.—Dr. W. A. Sturge [*] explains the pectoral pain as well as the associated dyspnœa, gastric disturbance, brachial soreness, etc., by the conversion of an efferent impulse conveyed along certain nerve trunks (the cardiac nerves) into a general commotion of a group of spinal nerve centers with which

---

[*] N. Y. *Med. Journal*, May 26, 1883.

these trunks are connected, and by the consequent involvement of other trunks springing from the same centres. So in visceral inflammations he believes we relieve pain by *counter irritation*, the latter exhausting the sensitiveness of the centre which supplies nervous force both to the viscus and the overlying integuments, so that by making a powerful impression on the latter we inhibit the irritation of the centre upon the former.

Nitrous Compounds in Angina Pectoris.—Dr. Hay,* after experimental observations, concludes that nitrous acid in any combination, whether as an ether or metallic salt, is useful in the treatment of angina pectoris, and that in the case of the nitrite of amyl the action of the acid is aided by that of the base. But all the compounds of nitric acid are useless unless they are so composed that the acid is converted into nitrous acid in the system. So far it appears that nitrogen containing remedies for angina pectoris may be divided into two classes, the one consisting of combinations of nitrous acid with metallic oxides or alcoholic radicles, the other comprising a peculiar class of nitric ethers obtained from higher alcohols. In both classes the action of the compound is ultimately on the nitrous acid present. Examples of the first class are the nitrite of sodium, and the nitrite of ethyl, and of the second class nitro-glycerine. Then there are the compounds of amyl whose action is similar to that of nitrites. But they are limited at present, the dose required is large, the action is not rapidly produced; and disagreeable after effects are apt to be produced. They are not to be chosen in the treatment of angina pectoris.

Nitro-Glycerine is a vaso-motor paralyzer and dilator of the peripheral vessels, and is of service in diseases of the heart, especially those of the aorta. It opposes the symp-

---

* The *Medical Record*, July 7, 1883.

toms of cerebral anæmia, as Dujardin Beaumetz* has shown to be the case with the nitrite of amyl. In cardiac affections characterized by feeble state of the myocardium amyl nitrite has been regarded as a cardiac stimulus, and nitro-glycerine has doubtless the same action. Nitroglycerine has been especially productive of good results in angina pectoris in doses of three drops a day of a solution of one to a hundred. It has also been successfully used in tendency to syncope and in palpitations, but only in those cases in which there is a nervous or anæmic state. Free diuresis has been produced. Formula: Distilled water, fl ℥ x, solution of nitro-glycerine—1-100, gr. xxx; three teaspoonfuls a day, one after each meal.

Angina Pectoris and Nitro-Glycerine.—Dr. Jacob Frank† of Buffalo treated a woman aged thirty-one years with morphine, electricity, etc., with little benefit. He then tried nitro-glycerine m. j. of one per cent solution, gradually increased m. vj. three times a day and then decreased. The patient had no more of the disease. After she had been taking the medicine for six weeks, aphthæ appeared on the tongue, mouth, and fauces. The medicine was stopped, and these ulcers were treated with borax, etc., after which the treatment was resumed.

---

* Dr. Henri Huchard *Am. Jour. Med. Science*, July, 1883.
† The *Medical Record*, May 5, 1883.

LECTURE XV.

DEFORMITIES OF THE HEART.

CERTAIN congenital malformations of the heart take place, the result of imperfect fœtal evolution; to enter into a lengthy discussion of these conditions would involve an extended inquiry into the science of embryology, which is not intended. The principal deformities to which the organ is subject may be briefly summarized, as arrests of development during some period of intrauterine existence, in which the heart may either consist of two or three cavities: those in which, although four cavities exist, there is a deficiency in the septa between the right and left sides, and those cases in which the heart is not confined to its proper cavity. These conditions may be best elucidated by the following cases.

Disease of Pulmonary Valve and Open Septum.—Dr. Bruen,* in the Pathological Society of Philadelphia, describes a heart thus: Two of the semilunar leaflets of the pulmonary artery are nearly destroyed by atheromatous changes; the third segment is much thickened and projects as a leaflike fold, roughening the mouth of the pulmonary artery. This vessel is dilated to twice its natural size, forming nearly an aneurism. The vessel's walls are covered with vegetations of inflammatory origin, or due to atheromatous changes. The right auricle is very small and imperfectly developed, the bulk of the cavity being formed by the auricular appendix. The tricuspid valves are much thickened, but probably competent. Between

---

* *Medical News*, Dec. 23, 1882.

the ventricles is an orifice that will admit the forefinger. It is directly beneath one of the tricuspid leaflets and is lined with endocardium. The walls of the right ventricle are thinned and its cavity somewhat dilated. Dr. Bruen ascribed the perforation to the pressure of blood in the right ventricle. The dilated artery made a pulsating tumor to the left of the sternum, between the second and fourth ribs. Over the tumor a "post diastolic and presystolic bruit-like murmur could be heard," while close to the junction of the second and fourth ribs to the sternum a hoarse systolic murmur could be heard. There was no cyanosis.

Dr. Wilson thought that in this case the left heart by its greater strength had sent its blood into the right and overloaded it, and that the condition of the pulmonary artery was due to this.

Open Ductus Arteriosus.—Dr. Beverly Livingston* reported to the N. Y. Pathological Society a case of this kind. In a newborn male child there was no indication of disease till the morning of the third day. Then there was a sudden arrest of respiration; an irregular action of the heart, cyanosis, first of the face, then of the whole surface of the body. In a minute and a half the pulse resumed its regularity, and the skin its natural color. At intervals of about an hour the same symptoms recurred, and as they recurred they lasted a little longer, about two minutes, and artificial respiration was necessary for the restoration. In the sixth attack the best efforts were unavailing. The heart ceased beating forever.

The lungs showed numerous small regions of collapse (atalectasis), and the right heart was distended with blood.

---

* Path. Soc., June 27, 1883.

Open Foramen Ovale.—Dr. Ferguson exhibited at a meeting of the N. Y. Pathological Society two hearts illustrating persistent foramen ovale. One opening admitted a tube half an inch in diameter. Dr. Allchin of London asked Dr. Ferguson why he supposed the blood in one of his specimens passed from one auricle to the other. He had seen the lesion very frequently, and supposed that it was generally accepted that the pressure of the blood in both cavities was so nearly equal that it kept the opening sufficiently closed to prevent any evidence that the blood passed from one auricle to the other.

Dr. Ferguson replied that the arrangement of the tissues was such in one of the cases that he thought pressure on one side would close the foramen, and that pressure on the other would open it.

Ductus Arteriosus Open in a Dog.—Dr. W. H. Howell and Professor Donaldson* found in a dog the pulse rate greater and the cardiac impulse stronger than normal. The apex was further to the left than it should be, showing marked hypertrophy. Over all the cardiac region Prof. Donaldson heard a loud rasping cardiac murmur with maximum intensity over the base, also a slight murmur with the second sound. Experiments showed a normal arterial pressure and no abnormality in the sphygmographic tracings except an unusually marked dicrotism. The ductus arteriosus was found open, its calibre being at least equal to that of either of the pulmonary arteries. With such free communication between the aortic and pulmonary circulation it is noticeable that so good an average arterial pressure was maintained (as measured in the femoral artery). Only one such case is found on record.

I have not the "circular" in which this account was

---

*N. Y. *Med. Jour.*, March 3, 1883.

first published, but this abstract, as I suppose it is, leaves the impression that "the loud rasping murmur" was produced by blood flowing through the open ductus arteriosus. But there is no reason to suppose that the ductus arteriosus produces any more noise when the blood passes through it than does the aorta, or the carotid where it divides into internal and external. It is easy enough to hear the sounds of the heart in utero, yet "a loud rasping murmur" is almost never heard. The report does not mention the aortic valves.

Then the blood pressure was apparently taken in the femoral artery, where it would represent the strength of the left ventricle, and a good part of that of the right, and until the heart became enfeebled, the pressure ought to be considerable, at least compared with that in the carotid artery.

Congenital Communication between the Right Side of the Heart and the Aorta.—Dr. B. Livingston* exhibited to the N. Y. Pathological Society the heart of a child, a male three months and twenty seven days old, that had been strong and healthy, and had never given any symptom of heart disease. He died of entero-colitis.

Thymus gland large; heart enlarged; at root of the aorta on the right side was a communication with the right heart. "Above and below the tricuspid valve it was closed by a part of this valve, and most likely no blood passed through it." The valves of the aorta were only two in number, but were sufficient to close the opening. The mitral and tricuspid valves were normal. On the mitral were plenty of those semi-transparent nodules referred to in other cases.

Congenital Pulmonary Stenosis.—Dr. B. Livingston†

---

* *Med. Record*, Sept. 1, 1883.
† Ibid.

reported to the N. Y. Pathological Society an instance of this obstruction that occurred in a female child one year and eight months old. The parents believed that the disease existed at birth. When Dr. Livingston first saw her she was pale, not cyanotic, short of breath, disposed to rest. A systolic very loud murmur was heard most distinctly at the union of the fourth left rib with the sternum, and nearly as distinctly at apex; was also heard in sixth intercostal space to left of apex beat and between the scapulæ. There was a distinct fremitus over the præcordial region felt by the hand. She died of septicæmia following mild ulcerative stomatitis.

"The three semilunar valves of the pulmonary artery had grown together, so as to leave a round hole two millimetres in diameter, and on one side there was a small vegetation, which must have been an obstacle to the current of blood. The foramen of Botal was enlarged and open, and the interauricular septum was very much thinned, and there were some small holes through it, so that it had a cribriform appearance." The heart was hypertrophied and the left side dilated.

Stenosis and Atresia of the Pulmonary Artery and Opening in the Ventricular Septum.*—A boy of thirteen years. He was under observation about four and a half months. In the boy's opinion and memory he had been cyanotic from four years of age (probably from both). He had attacks of deep cyanosis lasting some days, during which he would become wholly or partly unconscious, the respiration slow and sighing, the pulse rapid and feeble, 120 to 150, the pupils widely dilated and insensible to light; the temperature at first somewhat elevated, 100° to 102°, but later it would fall slightly below the standard, while the extremities would be cold and covered

---

* Dr. B. A. Watson, *N. Y. Path. Soc. Med. Record*, March 10, 1883.

by a slight perspiration. These attacks were generally produced by slight fatigue, or exposure to the sun (it was in summer) and were always preceded by several hours with severe headache and a general feeling of illness. He died in such an attack in which for the first time he complained of a severe pain in the præcordial region.

Two murmurs were heard at the base and two at the apex. During the attacks only, subscrepitant râles were heard over the whole chest.

The heart was found to be large; the left auricle was only half the capacity of the right; the tricuspid and mitral valves were healthy. Two segments of the aortic valve were sealed together by a material of cartilaginous firmness; the valve was atheromatous and covered with vegetations. There was marked stenosis of the pulmonary valve. There was an opening through the ventricular septum which permitted the little finger to pass through it easily. The left carotid artery arose from the innominata and the left subclavian nearer the median line than usual. It is not probable that there were two murmurs at the apex distinct from those at the base, but it is probable that the base murmurs were conducted so strongly as to deceive the observer. It seems that no murmur was recognized that could be referred to the opening in the septum.

Dr. Eskridge exhibited, at the meeting of the Pathological Society of Philadelphia, March 3, 1883, an unusual deformity of the aorta and the aortic valve. The specimen was taken from a man about seventy years of age, who had suffered a number of years from severe heart disease. The walls of the large arteries were thickened, rigid and contained numerous deposits of inorganic matter. The left ventricle was enormously enlarged. The first half inch of the aorta was hard and unyielding from fibrous thickening and calcification. Each section of the valve was almost wholly transformed into a bone-like matter, and at points the valves were a quarter of an inch in thickness. They were immovable and almost entirely closed the opening. "One of the leaflets, about three fourths of an inch in all directions, with its vegetations stretched across the aorta, lay against, and was apparently adherent to the other segments of the valve, the latter being curled upon themselves. The central portion of the aorta was entirely occluded, and only two small openings remained" between the segments of this valve near the artery:—one three m. m. by one, the other about two thirds this size. Then other similar openings had existed, but were obliterated before death by a thin fibrous transparent membrane. The valves on the aortic side were very rough, but on cardiac side were rather smooth. A vegetation ten m. m. long was attached to one of the segments. One segment was adherent to aorta for about half an inch, while the free portion was folded on itself and pointed toward the orifice.

Dr. H. Horace Grant* reports the case of a mulatto girl aged 16, who was poorly developed, had never menstruated, had often coughed up blood, had remarkable clubbed fingers and toes, had a great deal of dyspnœa, had a loud tricuspid regurgitant murmur, and increased præcordial dulness. Her ill-health was traced back to her birth. Her complexion forbade the recognition of cyanosis.

The right auricle was greatly dilated, the right ventricle moderately, the tricuspid valve was insufficient. There was no pulmonary artery. The aorta was given off just above the septum, one half from the left ventricle and one half from the right. The three sections of the aortic valve arose, one on the right, one on the left of the aorta,

---

* *Am. Jour. of Med. Sci.*, July, 1883.

and the third, posterior, from the top of the septum. The two coronary arteries arose from the right sinus of Valsalva. The pulmonary arteries were given off from the front of the aorta at the pericardial attachment each about half an inch in diameter.

Sundrifold Pulmonary Valve.—A young man 20 years of age died from fracture of the skull caused by a fall down a flight of stairs when intoxicated. The valves of the pulmonary artery were tested by water and were found slightly incompetent. There were four valves (or cusps); three of about the ordinary size, the fourth much smaller than the others, and imperfectly separated from one of them. The other valves of the heart were healthy, and the heart was of natural size.

He had been examined by Dr. Begbie repeatedly during the three years preceding his death. A very decided thrill and a loud blowing murmur attended the systole at the left border of the sternum at the third rib. The thrill was very limited. The murmur was diffused over the upper part of the chest, but was scarcely appreciable in the carotids. There was a much fainter diastolic murmur almost limited to the same spot. (Begbie's Works, Sydenham.)

Double Ventricle and Insufficient Septum.—Mr. Stone, in St. Thomas's Hospital Reports,* gives the following: Heart not excessively large, the vessels given off normally. Ductus arteriosus closed; water injected by the aorta comes out freely by the pulmonary artery. The finger passed into the pulmonary artery meets obstruction an inch and a half below the valve. The auricles communicated by a slit-like fissure such as is not uncommon without producing any pathological effect. The walls of the right ventricle were hypertrophied to exactly

---

* *Am. Jour. of Med. Sci.*, Oct., 1883.

an equal in thickness to those of the left. The cavity of the ventricle was divided into two chambers, one much smaller than the other, and almost completely shut off from it by a firm fleshy partition. These were in communication with each other by a small circular aperture with cartilaginous margin, studded with vegetations of the size of a millet seed, about a quarter of an inch in diameter. The small oval chamber was an inch and a half long, situated between the general cavity and the pulmonary valve. This was quite healthy. The septum between the ventricles was perforated by a large semilunar orifice in its upper space.

During life there was a distinct systolic thrill over the cardiac region, most marked at a part half way between the left mamma and the sternum, and conveyed upward in a diagonal line from the midsternal toward the outer extremity of the left clavicle. This was accompanied by a loud, rough sound, also systolic, most accentuated at the point covering the origin of the pulmonary artery. It was not loud at the apex of the heart, was also lost to the right of the sternum, but was audible over the upper part of the scapula posteriorly, and less distinct lower down. The heart was somewhat enlarged toward the left side. "The pulmonary valve was healthy," and as nothing is said of the other valves it is wrong to infer that they were healthy also. The thrill and rough sound must have been produced at "the small circular aperture."

No Aortic Orifice.—Dr. Alfred Meyer[*] reports the following case : A male infant twenty-four days old, cyanotic all over the body, suffering great dyspnœa, no radial pulse on either side, distinct pulse in the dorsalis pedis, 135. Nothing abnormal was found in the lungs, but there was

---

[*] The *Med. Record*, April 21, 1883.

a loud, distinct blowing systolic murmur in the neighborhood of the second and third ribs to the right of the sternum. The mother said he had been short of breath from his birth, but that the skin was not discolored till the third week. The gradual increase of the dyspnœa while nursing and the deepening cyanosis alarmed her. The child was found dead in bed forty-eight hours later.

The pericardium contained three fourths of a drachm of clear serum; heart substance of dark blue color. The right auricle and ventricle greatly dilated; foramen ovale pervious; pulmonary artery fully twice its normal size; ductus asteriosus pervious; left auricle one third the size of the right; walls of the left ventricle half an inch thick; its cavity just large enough to hold a small pea; pectinean muscles fused into a solid mass; mere traces of the chordæ tendineæ. *Aortic orifice completely closed;* dictus arteriosus leads into the arch of the aorta, which is about one eighth of an inch in diameter and ends in a blind sac at the base of the heart. The septum ventriculorum complete.

This patient lived twenty-seven days. Two other recorded cases give a life of two and four days respectively.

Atresia and stenosis of the aorta at or above the opening of the ductus arteriosus is comparatively common. Eppinger collected 42 cases and adds 2 of his own. The patient may live 50 or 60 years; one lived to 92.

Both Septa of Heart Deficient.—Dr. H. A. Gallatin[*] records the case of a male infant thirty months old, in whom the heart was as large as that of an adult female, the increase almost wholly in the right cavities, the auricle and ventricle having each a capacity of two ounces, and the walls of the latter were half an inch thick; a probe passed through the valves of the pulmonary artery appeared

---

[*] The *Medical Record*, Oct. 20, 1883.

in the aorta. There was an opening in the ventricular septum at the top three quarters of an inch in diameter, so that the aorta and pulmonary artery both communicated freely with either ventricle, or with both, most freely with the right. The semilunar valves of the aorta and pulmonary artery were, however, perfect. The foramen ovale was open three quarters of an inch. The Eustachean valve had disappeared.

The patient was tall and delicate, upper parts of the body usually œdematous, subject to turns of cyanosis, and even fainting; surface cold and patient chilly; there was dyspnœa by turns. Varying murmurs not constant; the most constant was a presystolic *rippling* sound.

A Diverticulum, or Glove-Finger Extension of the Left Ventricle.—Dr. Gilbert* records the case of an infant who lived to the age of ten months in whom was a ventral hernia and defective abdominal wall in the middle line from the umbilicus upward. The diaphragm was found to be defective also, so that the pericardium opened into the abdominal cavity; and projecting into the abdomen was a pouch-like diverticulum attached to the apex of the left ventricle, with muscular walls and having internally fleshy columns like the ventricular cavity itself. This appendix was thirty-eight millimetres in length, and shaped like the finger of a glove. Dr. Peacock in his "Malformations of the Heart," though he refers to several cases of defect of the pericardium, does not mention any such abnormality.

Dextro-cardia, the other Viscera in their Proper Places.— Dr. D. Frank Sydston† reports the case of a man twenty-four years of age, a barber. He says: "The heart occupies precisely the same relation on the right side that it

---

* *Med. Record*, Aug. 11, 1883.
† Ibid., July 21, 1883.

ordinarily does on the left, the apex being in the right fifth intercostal space." Otherwise the heart appears to be perfectly healthy. "The abdominal viscera are in their normal positions, a point to which especial attention is directed, inasmuch as such cases are usually accompanied by a transposition of the liver and spleen." He refers to the pathological displacements of the heart of which this is not one, as he has none, and has had none of the diseases which cause such displacement. He calls it, therefore, "An instance of a quite rare anomaly, congenital dextrocardia.

Open Foramen Ovale; Perforated Septum Ventriculorum. —Reported by Dr. Toupet.* A child, seven years old, had had measles and whooping cough; cyanosis from birth, blue, temperature, 96.5°; eyes prominent; nails enormously hypertrophied; heart very large; respiration rapid and labored; at the level of the third costal cartilage, near the left border of the sternum a loud blowing systolic sound. On inspection the heart lay nearly in the median line owing to the greater size of right ventricle, right auricle normal, but left rudimentary; aorta greatly dilated up to the origin of the left subclavian artery, beyond that, normal. The pulmonary artery only one fourth the size of the aorta; the walls of the right ventricle very thick, tricuspid valve healthy; the infundibulum separated by a partition from the right cavity, which was pierced by an opening of the size of a small goose quill; the mitral valve was normal; in the upper part of the ventricular septum was a hole that admitted a finger; the aorta arose exactly at the septum, communicating with both ventricles; valve of the foramen ovale was not adherent and was pierced by two openings of the size of a crow-quill; the lungs were congested.

---

* *Med. Record*, Aug. 4, 1883.

Malformation (?) Aortic.—Dr. Thomas B. Peacock* gives the following description of the condition of a heart and thinks the lesion was congenital. "The right and posterior segments of the aortic valve were blended together so that the aortic opening had but two valves; both of them were very much thickened, and the under curtain fell below the level of the other curtain, so that there was obstruction and regurgitation.

The action of the heart was tumultuous and visible over a large space; there was decided prominence in the præcordial region. Dulness on percussion began in the second interspace and became entire in the third. Laterally it begins to the right of the sternum and extends beyond the line of the left nipple. At the base there was a systolic murmur heard most distinctly at the right side and upper part of the sternum; it was short and rough and followed by a soft diastolic murmur which was propagated down the course of the sternum. Toward the apex there was a creaky murmur which was independent and perhaps presystolic, not heard posteriorly. Perhaps a slight passing tremor felt at the apex.

Ectopia Cordis.—M. Tarnier† exhibited at a meeting of the Académie de Médicin a woman whose sternum was bifurcated at the lower part, so that the beating of the heart could be seen to take place immediately under the skin of the epigastric region. The ventricular part could be seized between the fingers; by palpation over the upper part of the notch the contractions of the auricles could be detected. Apparently the diaphragm did not exist under the heart. M. Beau believed that the beat of the heart was due to dilatation under the influence of the afflux of blood at the time of the

---

† *Am. Jour. of Med. Sci.*, Oct., 1883.
* The *Medical Record*, Oct. 6, 1883.

ventricular systole. According to him the apex of the heart contracts during the diastole. In this case it was easy to perceive that the ventricle was soft during diastole, and hard during the systole. In systole the apex of the heart struck the thoracic wall.

LECTURE XVI.

FUNCTIONAL DISEASES OF THE HEART.

Forced Heart.—There is a diseased condition of the heart observed for the most part in those who have made great and prolonged physical exertion, whether in violent games, heavy labor, climbing ladders, running matches or forced marches, heavy drills, and the like, which has been called "the overwrought heart," "the forced heart," the irritable heart.

This condition most physicians have seen, and perhaps have not distinguished it from irritable heart arising from other causes. It has been described in France and Germany, and in England by Maclean and Myers. Cases have been seen in fifties abroad, but as our war brought into conflict a greater number of men than were ever mustered into the field before, the medical muster should be correspondingly large. Dr. Da Costa alone, in a military hospital under his charge, saw more than three hundred cases. He had the means, therefore, of giving, and probably has given, the best description of the disease which we possess. It may be found in the *Am. Jour. of Med. Sci.*, Jan., 1871. I also had large numbers of army patients in the years of the war. But as Bellevue Hospital stands on the river, and the helpless cases had to be carried on stretchers only about two hundred feet, they were all landed there, while the walking cases were sent to hospitals more distant from the river. These cases of irritable heart are almost all walking cases, and therefore passed by us to the upper hospitals.

I am not ignorant of this disorder, but am about as familiar with rupture of the valves or heart wall from strain as with the "overworked heart." It is wise then to take the portraiture of the disease from Dr. Da Costa, whose opportunity of studying it has so far excelled that of anybody else.

Dr. Da Costa says in substance that a soldier in active duty is debilitated by a diarrhœa or fever, but soon cured, returns to duty, but finds that as he exerts himself he gets out of breath, and falls behind his comrades in the march; has dizziness and palpitations of the heart, and pain in the chest. He is oppressed by his gun and knapsack. At the hospital his rapid pulse proves his infirmity, although he looks like a man in sound health. Any associated disorders may soon pass away, but the irritability of the heart, the excited organ only slowly returning to its healthy action, or the excitement may continue till the patient is discharged from the army or is sent to the invalid corps.

There are, he says, many others in whom the irregularity of the heart's action and pain in the region of the heart occur more suddenly without previous disease.

Dr. Da Costa describes the symptoms, and says that in those not having organic disease of the heart they were as follows:

Palpitations.—In some the attacks lasted many hours. They occurred at all hours of the day or night, and were repeated in some six or more, in others occurring but once in twenty-four hours, while some were free from them for days. Sometimes the attacks were so violent as to produce unconsciousness, while physical exertion was a uniform cause of palpitation; they occurred also when the patient was in bed and even asleep. Some were worse at night and early morning. The attacks were accompanied by cardiac uneasiness and pain, and pain under the left shoulder and in some by headache, dimness of vision, and giddiness. With some exceptions the patients would not lie on the left side in bed, fearing that that position would provoke an attack. But there were some who preferred the left decubitus, and some who could lie on either side or on the back.

Pain.—This was almost a constant symptom, varying in character from a dull ache to sharp and lancinating. In a few it was a burning sensation, or tearing. Sometimes it preceded the palpitations, and sometimes occurred without them. In rare instances it was relieved by muscular exertion, but was generally caused by it. Deep breathing in some increased the severity of the pain.

The chief seat of the pain was over the heart near its apex; it was also felt above the inferior angle of the left scapula, in the left axilla, and down the left arm, attended there with numbness. Sometimes it radiated from the heart in every direction. With the pain there was hyperæsthesia in the cardiac region. The pain, the author thinks, was not an intercostal neuralgia, but that affection sometimes occurred as a complication.

Pulse was generally very rapid—100 to 140—small and compressible. It might or might not have the jerky character of the heart beat, and was subject to great variations. It was noticed the change in frequency was greatly affected by change of position. In one case, the patient standing, it was 105 to 108, after lying down less than 80, and fuller, in another it changed from 124 to 94. These pulse records were made during the absence of palpitations; during these the frequency was increased. The author also observed a frequent tendency to cyanosis.

Respiration.—Respiratory oppression on exertion and in the palpitations was constant; a little of it was felt at all times, and there were occasions when the patient could not lie down. The normal ratio of the breathing and heart beats was quite broken up. A man with a pulse at 124 breathed 25 times a minute, one with a pulse at 146 breathed 26 times, and one whose pulse beat 192 times breathed 26 times.

Nervous Disorders.—The headache was constant as a symptom, or in duration. It is most likely to follow severe palpitations. Dizziness was not uncommon. It was increased by exercise and stooping; often preceded the palpitations, and in one case at least caused a fall from a horse. Disturbed sleep, jerkings in sleep, and frightful dreams were not uncommon.

Itchings of the skin and excessive perspirations, or of the hand only, were complained of by several.

Digestive Disorders, with "irritable heart," indigestion, abdominal distention, and diarrhœa are very common, but Dr. Da Costa regards them as more connected with the cause of the disease than a consequence of it.

The Urine contained oxalate of lime and other products of impaired digestion sometimes, but never anything that was peculiar to this affection.

Physical Examination.—The beat of the heart was quick, and not remarkably strong, often irregularity of rhythm. The first sound of the heart is feeble and short, resembling the second in duration. The second sound is sometimes increased, always distinct. Sometimes, he says, "the sounds of the heart are split," "there were double beats and intermissions."

The disease, for the most part, either slowly subsided or was followed by degrees, by enlargement.

Dr. Da Costa, referring to causes, has constructed the following table, showing its relation to fevers and diarrhœa. The symptoms did not generally appear till the patients were sent back to duty.

200 CASES.	P. C.
Fevers	17.
Diarrhœa	30.5
Hard Field Service	38.5
Wounds, Injuries, Rheumatism, Scurvy, Soldier Life, and Doubtful	18.
	100.

He has another table in which the "results" are shown in 200 cases, and in which death finds no place.

Treatment.—In Dr. Da Costa's remedies *rest* has the first place—rest in bed, not only at night, but for many hours of the day. His most esteemed drug is digitalis. He gives the next place to veratrum viride. Aconite found but little favor with him; gelsemium none at all. Belladonna corrected the irregularities of the heart, but did not reduce the frequent beat. Opium he used only for the diarrhœa and severe pain, but he declined to use it against the irritability, fearing that in a long use of it he would establish the opium "habit." Hyoscyamus while not valueless, was not much praised. Conium, cannabis indica, valerian, ergot, and the bromides did not win his confidence to any great degree.

Tonics were often valuable, and zinc had some control over the disease. Iron was valuable in the anæmic cases.

Dr. Da Costa's paper is one of the most valuable of the age. It is a structure built by square and plummet.

Heart Sympathies.—The heart bounds with joy, it sinks in great disappointments, at sights of horror. Some persons faint at the sight of blood. Fear does not often paralyze the heart, as it often does the muscular strength, but it causes rapid and small pulse, while it contracts the capillaries of the surface and brings paleness and nervous tremors or large shakings. A gentleman was brought to me by the physician of a life insurance company for my opinion as to the freedom of his lungs from phthisis. On the first examination, for some reason, now forgotten, I was not willing to answer the question: should the risk be taken? On a second examination, I answered affirmatively. This fixed the gentleman's attention strongly on this liability to consumption. Two or three months afterward a messenger summoned me to visit him at once. I replied, "Call his physician" (the late Dr. Van Buren). The messenger soon returned stating that Dr. Van Buren was not at home and urged that I must go at once. It was raining a deluge, but I waded to the gentleman's house. He was on a sofa, was as pale, or rather white, as the pillow on which his head rested; his movements were as tremulous as if he had the "shaking palsy." His pulse was small and rapid. His voice was low and on a monotone. I asked what had happened. He said he had been *raising blood*. I asked him, How much? An attendant raised and unfolded a handkerchief which had on it numerous stains of blood. The blood had spread into and was diffused in the threads of the cloth, not in coagulated masses with slight diffusion, as blood from the lungs would be. It was blood that was mingled with the saliva, which had delayed its coagulation. I asked him if he had hurt his tongue while at dinner. He did not know that he had. I asked for a candle. His wife brought it in a hand trembling hardly less than her husband's. I found a laceration, a penetrating wound in the left side and back part of the tongue, as if a woodsplinter or the fragment of a nail from a flour barrel had been forced by a molar tooth into the tongue. I informed him of my discovery, and then he remembered that he did have a pain in the tongue while at dinner. But when he was informed that the blood came from the tongue and not from the lungs, he exclaimed: "Why, wife, I feel better. Why, I can sit up, I can stand up, I can walk," which he did then and there across the floor. I returned through the rain to my dinner and heard no more from my patient. The load of fear lifted from his heart, it resumed its natural mode of working, and the vaso-motor nerves relaxed their grasp on the capillaries.

Cardiac Neurasthenia.*—In some cases of exhaustion from continuous over-work, the symptoms centre chiefly about the heart. The symptoms are feeble cardiac action, giddiness, weakness, intermittent beat. There may be palpitations, dyspnœa, and even syncope. A physician who suffered in this way writes that he was relieved entirely by the following prescription:

℞ Quin sulphate..gr. xxiv.
    Mist camph...ad. ℨ vj.
    Acid hydrobromic dil............................... ℨ iij.
    Tinct digital... ℨ ss.
    Liq. aurant...ℨ j.
    Tinct. Nuc vom...................................... ℨ ij.
M. Half an ounce three times a day.

Cardiac Typhoid.—Mr. Bernheim† designates by this term cases in which, without notable organic alteration of, without pulmonary complication, or others capable of explaining the occurrence, the pulse becomes small, frequent and depressed. The patient succumbs to this paralytic acceleration of the heart, which may occur at the beginning of the fever, with or without concomitant nervous adynamia or at a more or less advanced stage of it. The temperature may be moderately febrile, normal or

---

* The *Medical Record*, Jan. 20, 1883.
† Ibid., Jan. 20, 1883.

subnormal. He considers this nervous asystole in typhoid fever to be due to the direct action of the poison on the cardiac innervation. The author bases his conclusions on six cases with autopsies. They ended in sudden death and no alteration of the heart correspondent.

Cardiac Vertigo.* I cannot say I am very familiar with, aside from that faintness which may attend grave aortic obstructions and regurgitations. The *Medical Record*, however, quotes Germain See as saying that when it occurs in the course of heart disease, we need to look no further for its cause, as a rule; that in cases of troublesome vertigo the heart should always be examined for its cause, especially if extreme pallor, a prominent symptom of aortic insufficiency, be present, and will not yield to medication... In these cases there are also the characteristic pains of angina pectoris, coming on every few months— another prominent, though not constant, symptom of aortic incompetency.

LECTURE XVII.

THE EFFECTS OF CERTAIN DRUGS ON THE HEART.

Accidents Caused by Tobacco.—Vallin† has lately reported to the Société de Médicine Publique some cases of poisoning by tobacco in smokers that are very emphatic. In M. Vallin's cases symptoms akin to those of angina pectoris were produced. Indeed, he calls the affection angina pectoris. Take a case. It is a "young lieutenant of vigorous constitution who for a year had been subject to attacks of angina pectoris. The

---

* *Med. Rec.*, Sept. 22, 1883.
† *Annales de Hygiene Publique*, April, 1883.

attacks, at first rare, had become more frequent—indeed, almost daily. The day after his admission to the hospital I witnessed an attack—atrocious retro-sternal pains, with painful engorgement of the left side of the neck, extreme anguish, paleness of the face, cold sweat, a tendency to fainting, respiration deep and sighing, a marked slowness of the pulse, fifty-two pulsations with irregularity and intermittence. This state of præcordial anxiety continued about twenty minutes. The patient often said he was about to die. At length the pains ceased; the face regained its usual color. There remained a tendency to vertigo when he was in the erect position, which continued for half an hour more." This young man had been admonished that he must stop smoking, and he had adopted the advice, but his friends assembled in his room every evening, and smoked from eight to eleven o'clock. M. Vallin informed the patient that he could as well be poisoned by the smoke of others as by his own. He appreciated this information, and from the next day these attacks grew milder, and after ten days there were none that were at all marked, and M. Vallin adds: "I have reason to believe that the suppression of the cause led to the definite suppression of the attacks."

Another case follows in which there was great dizziness and tendency to faint, a slow pulse, twenty-four a minute, with intermittent pulse and arrest of the heart beats, ringing in the ears. The attacks were preceded and accompanied by extreme præcordial anxiety without localized pains, but with suffocation and dyspnœa. The syncopal attack recurred about every two hours. The patient had had rheumatism, and M. Vallin, though there was no murmur or enlargement, could not but believe that these symptoms were the result of rheumatic changes in or on the heart till he learned that the patient had been a great smoker. Abstinence from tobacco had nearly effected a cure when M. Vallin, wishing to confirm his second diagnosis, allowed his patient a pretty strong segar. When he had smoked one half of it he was attacked with great præcordial anguish, a tendency to faint. He was obliged to go to bed, and to keep the horizontal position, and this time had great frequency of pulse. ·The same experiment was tried once more in a more advanced stage of convalescence with the same results, and this time there was great slowness of the pulse, and then followed great weakness of the legs, so that he could not walk for a week. His recovery, however, was complete in the end.

A physician makes his third case. This person smoked excessively. He had from time to time pretty strong intimations that this excess was injuring him. Nature seemed to assume the office of monitor by causing an absolute disgust for tobacco of several days' duration; still he smoked when he could and all he could, till symptoms of angina pectoris fully alarmed him. He then stopped smoking and got well.

M. Vallin refers to a dozen cases reported to the Academie des Sciences by M. Decaisne of intermittence of the heart and pulse caused by the abuse of tobacco. Also to a sort of epidemic of angina pectoris which occurred among sailors. There was a great tempest. They were confined in a very small apartment in which all ventilation was prevented to keep out the sea. They smoked excessively in this confined air. The men were attacked —the report does not state in what time—with this disorder both those who smoked and those who did not. This disorder ceased in a little time after by the suppression of tobacco. Some members of the society related analogous cases.

M. Vallin believes, and probably truly, that excessive smokers get at length a saturation of nicotine, and then a spark will set off the magazine, a slight cause will produce alarming symptoms, and that this susceptibility will last six months to two years.

Bad Effects of Nitrous Oxide on the Heart.—Dr. Ottley* observed the following: A young woman who had suffered from rheumatic fever which had left a slight mitral lesion, with a faint murmur, sometimes hard to hear, and a little enlargement of the heart, but no functional disturbances, took the gas for the extraction of a tooth. Nothing important followed. A few days later she took it again, and so much dyspnœa and cardiac irregularity was caused that the administration of the gas had to be suspended. Subsequently the patient suffered from palpitation and dyspnœa, the heart acted irregularly, and the murmur was very much louder. The heart now for the first time gave evidences of inadequacy. The case is interesting from its rarity, the gas having been given indiscriminately with surprisingly few accidents.

Action of Digitalin on the Heart.—Experiments on terrapins and frogs by Drs. H. H. Donaldson and M. Warfield.† They say when the heart is doing normal work digitalin decreases that work; that there is a rough relationship between the size of the dose and the extent of the decrease; that with small doses the pulse rate is at first increased; that the diminution of the heart's work is much more dependent on the strength of the dose at any given time than on the total amount of the drug administered. A large amount given in several hours has much less effect than a smaller amount given in a few minutes. Therefore, the theory that digitalin has a cumulative action lacks experimental confirmation.

Action of Alcohol on the Heart.—Prof. Martin‡ of the

---

* N. Y. *Med. Jour.*, June 16, 1883.
† Ibid., March 3, 1883.
‡ Ibid., Oct. 13, 1883.

Johns Hopkins University has found that alcohol in quantity not directly poisonous did not affect the rate of the pulsation. Blood containing one eighth per cent by volume of absolute alcohol did not affect the amount of work done. Blood containing one fourth per cent always diminished it, and might so reduce the amount sent out by the left ventricle that it was not sufficient to supply the coronary arteries. (About 0.46 of an ounce of alcohol for a man of 150 pounds.)

Alcohol, in Prof. Martin's opinion, does not act on the heart till it reaches the muscular and nervous tissues through the coronary arteries.

As alcoholic blood holds its oxygen more firmly than other blood, it is probable that the deleterious action of alcohol is due to the deprivation of oxygen which it causes in the tissues.

INDEX.

ACUTE pericarditis, serous effusion in, 41
Adhesion, pericardial, 41
Adonis vernalis, 206
Alcohol, effect of, 245
Anæmic murmurs, Graves on, 20
Angina pectoris, 216
" alcohol as a cause of, 216
" amyl nitrite in, 219
" chloral in, 218
" counter irritation in, 219
" digitalis in, 219
" electricity in, 219
" gouty, 216
" hysterical, 217
" hysterical, cases of, 218
" nitro-glycerine in, 219
" nitrous compounds in, 220
" morphine in, 218
" of organic origin, 217
" potassium bromide in, 219
" pathology of, 216
" tobacco as a cause of, 216
" treatment in the intervals, 219
" treatment of the paroxysm, 218
Aortic disease, 179
" malformation, 234
" obstruction, 181
" obstructive murmur, 13
" orifice, absence of, 230
" regurgitant murmur, 13
" regurgitation, pulse of, 185
" " second sound abolished in, 193
Asynchronism, 29
" from anæmia, 29
" from asphyxia, 29

Atheroma, 163
" pathological changes in, 164
Auscultatory percussion, 33

BALFOUR on anæmic murmurs, 18
Blood, course of, through heart, 10
Block on opening the pericardium, 90
Bony deposit in heart, 184
Bruit, cause of, 11
" de diable, 18
" de fluctuation, 94
" de la toupie, 18
" de pot fêlé, 94
" de rape, 12
" trumpet, 196

CACTUS grandiflorus, 213
Calcification of pericardium, 75
Cardiac and renal disease, 198
" disease, iron in, 206
" dilatation, Niemeyer on, 132
" disease, treatment of, 206
" irregularities, 29
" movements, interference with by adhesions, 42
" neurasthenia, 241
" œdema resulting from endocarditis, 168
" typhoid, 241
" typhoid fever, 121
" vertigo, 242
Carditis secondary to pericarditis, 46
Cook on double heart beat, 27
Convallaria majalis, 209
Coronary arteries, anastomoses of, 32

Cors bovinum, 123
DEFORMITIES of the heart, 222
Deformity of heart, 227, 228
Degeneration, adipose, 142
" fatty, recognition of, 143
" fibrous, 156
" pathology of fibrous, 156
" Quain's, 142
" Quain's, bicarbonate of soda in, 147
" Quain's, case of, 145
" Quain's, causes of, 146
" Quain's, diet in, 147
" Quain's, frequency of, 146
" Quain's, pathology of, 144
" Quain's, treatment of, 146
Depressor nerve, 31
Dextro-cardia, 232
Diastolic sound, 14
Differentiation of exo- and endo-cardial sounds, 47
Digitalin, effect of, 245
Digitalis, 212
" *infusion* of, 64
Dilatation, aneurismal, of heart, 190
" bleeding in, 140
" causes of, 132
" " " 136
" diagnosis of, 137
" frequency of simple, 137
" King on, 16
" of heart, 132
" Peabody on, 133
" physical signs of, 137
" rational signs of, 138
" simple, 136
" strychnia in, 141
" treatment of, 140
Direct murmurs, 13
Diseases of heart, formula for detection of valvular, 15
Disease, valvular, of right side of heart, detection of, 16
Double heart beat, 26
" " " explanation of, 28
Dropsy, caffein in cardiac, 206
" cardiac, Leech on treatment of, 214

Drugs, effects of, on heart, 242
Ductus arteriosus, case of open, 223, 224
Dyspnœa, causes of, 139

ECTOPIA cordis, 234
Effusion, position of pericardial, 40
Endocarditis, 96
" and hepatic abscess, 111
" diagnosis of, 104
" embolism in, 102
" extent of, 100
" in pyæmia, 111
" in rheumatic fever, 108
" Kuchler on fœtal, 193
" mechanism of valvular deformity from, 103
" murmurs in, 106
" pathology of, 101
" post mortem appearances of, 110
" prognosis of, 105
" statistics of, 148
" temperature curve of, 111
" thickening of valves from, 102
" treatment of, 111
" ulcerative, 112
" " 101
" " results of, 112
" vegetations from, 103
Endocardium, 9
Embolism, statistics of, 149
Extra-cardiac ganglia, 30

FATTY degeneration, 142
" " varieties of, 142
Fibrous degeneration, 156
Fibroid induration, 190
Fleshy columns, 10
Foramen ovale, open, 224
" " " 233
Forced heart, 235
" " pathology of, 235
" " symptoms of, 236, et. seq.
" " treatment of, 239
Francis on heart scanning, 26

Functional derangement, mitral regurgitation from, 213
Functional diseases, 235

GRAVES on anæmic murmurs, 20

HÆMOTO-PERICARDIUM, 94
Heart, bony deposit in, 166, 184
Heart clot, 159
" colors of, 160
" diagnosis of ante- and post-mortem, 159
" pathology of, 161
" when formed, 159
Heart, congenital opening between right side of, and aorta, 225
" deformities of the, 222
" disease of, due to rupture of tendinous cords, 14
" disease of right side of, 14
" disease, effect of, on cardiac nerves, 186
" hypertrophy of, 122
" intra-uterine disease of, 14
" largest reported, 133
" nerves of, 29
" position of normal, 39
" puncture and suture of, 89
" Roberts on puncture of, 89
" Westbrook on puncture of, 89
" rupture of, 150
" " from myocarditis, 150
" scanning, 26
" sounds, 9
" suture of, 155
" sympathies, 239
Hypertrophy, 122
" and palpitation, 131
" and pleurisy, 129
" causes of, 125
" concentric, 122
" definition of, 122
" diagnosis of, 128
" eccentric, 122
" from anger, 125
" from contracted kidney, 126
" from sexual excess, 126
" from valvular lesions 125
" pathology of, 123

Hypertrophy, physical signs of, 128
" possible extent of, 123
" rational signs of, 130
" simple, 122
" with dilatation, 122

INDIRECT murmurs, 13
Intra-cardiac ganglia, 30
Isham on the sphygmograph, 25

JUGULAR pulsation, 178

KEYT on transmission of pulse wave, 25
King on dilatation without injury, 16
Kuchler on fœtal endocarditis, 193

LEECH bites, results of enormous hæmorrhage from, 71
Leech on treatment of cardiac dropsy, 214

MICROCOCCI, 114
Mitral obstruction, consequences of, 191
" regurgitation from functional derangement, 213
" stenosis, 194
" " prognosis of, 192
" valve, 15
Murmurs, anæmic, 17, 18
" " Balfour on, 18
" constant diastolic, 100
" designation of, 17
" differential diagnosis of intra- and peri-cardial, 20
" in chlorosis, 18
" tones of, 17
" lying in, 196
" organic, disappearance of, 17
" pericardial, 20–21
" presystolic, 12
" tones of, 12
Myocarditis, 116
" cardiac aneurism in, 118
" causes of, 120
" consequences of, 120
" effusion in, 117
" embolism in, 117
" heart failure from, 121

Myocarditis, pathology, 116
" repair after, 117
" rupture in, 118
" symptoms of, 117
" symptoms of, 120

NERVES of heart, 29
Niemeyer on cardiac dilatation, 132
Nitrous oxide, effects of, 245

OBSTRUCTION, murmur of, 11
Œdema of valves, 104
Opium in uremia, 67

PARTZERSKY on drainage of pericardium, 92
Peabody on dilatation, 133
Pepper's case of tapping the pericardium, 82
Percussion, new method of, 34
Pericardial adhesion, 41
" adhesions, Gibson on, 88
" adhesions, signs of, 85
" drainage, 90
" effusion, 40
" effusions, encysted, 43
" incision, 90
" friction sound, 47
" friction, disappearance of, 51
" friction sound, double, 47
" friction sound; when heard, 47
Pericarditis, 32
" alkalies in, 60
" amount of effusion in, 54
" bleaching in, 63
" blisters in, 66
" cancer of œsophagus producing, 92
" causes of, 44
" citrate of potash in, 64
" clinical history of, 55
" connected with granular kidney, 44
" diagnosis of, 50 *et seq.*
" digitalis in, 64
" dry, 47
" following erysipelas, 44
" following hæmorrhage, 44

Pericarditis, following malignant disease, 44
" following pleurisy, 44
" following pneumonia, 44
" following rheumatism, 44
" following scarlatina, 44
" from contiguous inflammation, 45
" generally secondary, 44
" hæmorrhagic, 40
" mercury in, 59
" morbid anatomy of, 39
" occurrence in pneumonia, 57
" occurring in pyæmia, 45
" opium in, 66
" peculiar case of, 77
" physical diagnosis of, 46
" prevention of by diaphoresis, 65
" prognosis of, 56
" salicylic acid in, 61
" sero-purulent effusion in, 44
" sometimes primary, 44
" symptoms of, 46
" tonics in, 66
" treatment of 59
" walking case of, 55
Pericardium, aspiration of, 82
" Block on opening the, 90
" calcification of, 75
" indications for tapping, the, 80
" mode of puncture of, 82
" Pepper's case of tapping the, 82
" point of puncture of, 81
" Roberts on tapping the, 81
" structure of, 32
" tapping of; recovery, 84
" tapping the, 80
Pleurisy and heart disease, 192
Pneumo-pericardium, 92

INDEX.

Pneumo-pericardium, causes of, 92
" signs of, 93
Pulmonary valve, sundrifold, 229
Pulse-wave, rapidity of transmission of, 25

REGURGITATION, murmur of, 11
" without valvular disease, 13
Right ventricle, penetrating the, 91
Roberts on puncture of heart, 89
" on tapping the pericardium, 81
Rosenstein on pericardial drainage, 90
Rupture of the heart, 150
" " " causes of, 151
" " " from myocarditis, 150

SALICYL compounds, 213
Salicylate of soda, objections to, 63
Salter on cardiac rhythm, 26
Scabrous effusion, 41
Scarification, use of, 73
Second sound, accentuated, 23
" " exaggerated in aneurism, 23
Septa, both cardiac, deficient, 231
Septum, open, with disease of pulmonary valve, 222
Sibson on pericardial adhesions, 88
Sounds of heart, accentuated, 23
" " cause of, 21
" " mechanism of, 22
Solid stethoscope, 34
Sphygmograph, 24
" compound, 24
Stenosis, congenital, pulmonary, 225
" of right auriculo-ventricular opening, 193
" relative, 197
Suture of heart, 155

THROMBOSIS, cardiac, in acute disease, 161
Tobacco, effects of, 242 et seq.
Tonsilitis, murmurs in, 109
Tricuspid insufficiency, 191
" stenosis, 194
" valve, 14

VALVES, injuries to, 153–154
" " of heart, position of, 96
" pulmonary, diagnosis of disease of, 183
" rupture of, 172
" structure of, 9
Valvular disease, 163
" " causes of, 160
" " endocarditis as cause of, 170
" " general effects of, 175
" " of right side, 183
" " pathology of, 174
" " prognosis of, 204
" " relation of kidney disease to, 182
" lesions, duration of, 179
" stenosis, 173
Vaso-motor nerves, effect of action of, 31
Venesection, 69
Ventricle, diverticulum of, 232
" double, with insufficient septum, 229
" perforation of by gastric ulcer, 155
Ventricular septum, opening in, with atresia, 226

WET cups, use of, 73
West on pericardial drainage, 90
Westbrook on puncture of the heart, 89

www.ingramcontent.com/pod-product-compliance
Lightning Source LLC
Chambersburg PA
CBHW020803230426

43666CB00007B/835